Mathematics for Game Programming and Computer Graphics

Explore the essential mathematics for creating, rendering, and manipulating 3D virtual environments

Penny de Byl

BIRMINGHAM—MUMBAI

Mathematics for Game Programming and Computer Graphics

Copyright © 2022 Packt Publishing

Group Product Manager: Rohit Rajkumar
Publishing Product Manager: Nitin Nainani
Senior Editor: Hayden Edwards
Senior Content Development Editor: Rashi Dubey
Technical Editor: Saurabh Kadave
Copy Editor: Safis Editing
Project Coordinator: Sonam Pandey
Proofreader: Safis Editing
Indexer: Hemangini Bari
Production Designer: Roshan Kawale
Marketing Coordinator: Teny Thomas

First published: November 2022
Production reference: 1041122

Published by Packt Publishing Ltd.
Livery Place
35 Livery Street
Birmingham
B3 2PB, UK.

ISBN 978-1-80107-733-0
www.packt.com

Life-long learning has always been my goal. But I couldn't have done it alone. I'd like to thank Adrian who has been my friend, technical reviewer, and code tester throughout. He possesses the attention to detail I lack. And, I might add, if the code doesn't work – it's his fault. Next, I'd like to thank my wonderful daughter, Tabytha, who is in her first year of studying theoretical mathematics and computer science at university. She's been a great sounding board in helping me understand mathematical notations that I still struggle with. A mother could not be prouder than I am of her.

– Penny de Byl

Contributors

About the author

Penny de Byl is a full-stack developer with an honors degree in graphics and a PhD in artificial intelligence for games. She has a passion for teaching, and has been teaching games development and computer graphics for over 25 years at universities in Australia and Europe. Her best-selling textbooks, including *Holistic Game Development with Unity*, are used in over 100 institutions. She has won numerous awards for teaching, including an Australian Government Excellence in Teaching Award and the Unity Mobile Game Curriculum Competition. Her approach to teaching computer science and related fields is project-based, giving you hands-on workshops you can immediately get your teeth into. The full range of her teaching interests can be found at h3dlearn.com.

I'd like to thank the wonderful publishing crew and editors at Packt Publishing, who've been attentive, patient, highly professional, and available throughout the writing process.

About the reviewers

Amit Verma is a game developer and WebGL graphics engineer from Dehradun, Uttarakhand, who has developed web versions of the Pixlr photo editor. He has rich experience in the gaming industry, with development experience of more than 60 games and 30 WebGL applications, using Unity3D and WebGL. His passion and curiosity toward gaming technologies and their challenges have helped his clients to launch games on various platforms, such as web, mobile, and consoles. He holds a BTech degree in computer science and engineering from Uttar Pradesh Technical University, where he studied computer science and graphics programming. Currently, he is working at Excellarate as a lead WebGL graphics engineer for the development of metaverse content using WebGL.

I would like to thank my best friend, my wife, and the love of my life, Simran Gogna, for supporting me in my hobbies and professional career. Her selfless love and wise advice gave me the strength to reach my goals. I also wish to thank my mentors, Sumeet Arora and Shiwani Arora, for providing inspiration and direction in my early career days.

When **Soumya Mondal** is not saving the world, he cosplays as a student at the Technical University of Munich and as a visual AI researcher at a start-up. He has shipped a few games, raising awareness of social issues, and has a strong belief in the power of games as a medium. In his free time, he travels and runs.

Table of Contents

Part 1 – Essential Tools

1

Hello Graphics Window: You're On Your Way   3

2

Let's Start Drawing   21

3

Line Plotting Pixel by Pixel   41

4

Graphics and Game Engine Components   61

5

Let's Light It Up!   85

6

Updating and Drawing the Graphics Environment   99

7

Interactions with the Keyboard and Mouse for Dynamic Graphics Programs   113

Part 2 – Essential Trigonometry

8

Reviewing Our Knowledge of Triangles   141

Part 3 – Essential Transformations

12

Mastering Affine Transformations   223

13

Understanding the Importance of Matrices   243

14

Working with Coordinate Spaces   267

18

Customizing the Render Pipeline   359

19

Rendering Visual Realism Like a Pro   383

Preface

Mathematics is an essential skill when it comes to graphics and game development, particularly if you want to understand the generation of real-time computer graphics and the manipulation of objects and environments in more detail. Python, together with Pygame and PyOpenGL, provides the opportunity for today's developers to explore these features under the hood, revealing how computers generate and manipulate 3D environments.

Mathematics for Game Programming and Computer Graphics is a comprehensive guide to getting "back to the basics" of mathematics, using a series of problem-based, practical exercises to explore ideas around drawing graphic lines and shapes, applying vectors and vertices, constructing and rendering meshes, working with vertex shaders, and implementing physics techniques such as collisions and particle emitters. Using Python, Pygame, and PyOpenGL, you will create your own mathematical-based engine and API that will be used throughout to build applications and examples.

By the end of this book, you will have a thorough understanding of how essential mathematics is to creating, rendering, and manipulating 3D virtual environments and know the secrets behind today's top graphics and game engines.

When Packt first approached me to write this book, I had just released a course on computer graphics using Python and OpenGL in Udemy and on H3DLearn.com. The timing was never better than to review what I had learned in the course and write it up in a book focused on the mathematics involved.

Mathematics is one of those topics you either love, loathe, or have a quiet appreciation for. I was in the loathe camp for most of my university studies. I found it irritating and a time-consuming function I needed to get done to continue with my love of programming. Then, in my honors (fourth year), I was introduced to computer graphics and fractals. It was like a veil had been lifted for me and mathematics became magical, fascinating, and most importantly, visual. Being able to see the beauty in mathematics changed everything. I'm still not a lover of mathematics, but I've definitely gone from the loathe stage to quiet appreciation. Though I guess writing a book about it takes me from just appreciating it to love. You can't program any computer games or graphics without knowing the mathematics driving it all.

My goal in writing this book is to bring you into the "quiet appreciation" category. If you get to the "love" category, then great! I do believe that we all have different talents. Mathematics is not mine. I do find it quite challenging, but I have persevered and, over my career, learned to understand its origin and application. I am in no way one of those people who can multiply tens of numbers in my head, but I don't need to be. The point is that if you learn when and where to apply the mathematics, and then transform that into a programming algorithm, and you can validate and be confident of the output, then it doesn't matter whether you can calculate the output in your head or need time to work it out.

There's just so much mathematics to cover in this area that I could have honestly written another several hundred pages. However, it's my hope that the content herein will fire you up and give you the confidence and critical skills to independently further your education.

Who this book is for

This book is for programmers who want to better their 3D mathematics skills relating to computer graphics and computer games. Knowledge of high-school-level mathematics and a working understanding of an object-orientated language will be required.

What this book covers

Chapter 1, Hello Graphics Window: You're On Your Way, introduces the software tools used throughout the book to explore 3D graphics and game development, where Python and OpenGL will be used to develop graphical window applications.

Chapter 2, Let's Start Drawing, takes you through a series of exercises to set up and explore the basic skeleton code required to run and update a graphics application.

Chapter 3, Line Plotting Pixel by Pixel, contains a series of exercises to explore low-level line drawing algorithms, beginning with a naive approach that will reveal the issues involved in drawing lines on a raster display.

Chapter 4, Graphics and Game Engine Components, investigates the software architecture of games and graphics engines the associated data structures for working with 3D environments.

Chapter 5, Let's Light It Up!, shows you how to bring light and texture to a virtual scene to develop solid models of previously used meshes.

Chapter 6, Updating and Drawing the Graphics Environment, takes you through a series of exercises designed to reveal the cyclical nature of producing graphics frames in addition to building a strong foundation for your graphics project.

Chapter 7, Interactions with the Keyboard and Mouse for Dynamic Graphics Programs, teaches you how to use the Pygame API to gather user input via the mouse and keyboard and use it to interact with a game environment.

Chapter 8, Reviewing Our Knowledge of Triangles, reviews trigonometry with an aim to solidify the concepts of similar triangles and the mathematical properties of right-angle triangles.

Chapter 9, Practicing Vector Essentials, investigates the mathematical principles of vectors and explores the link between their properties and right-angle triangles to reveal the many ways they are useful in graphics.

Chapter 10, Getting Acquainted with Lines, Rays, and Normals, explores the similarities and differences between straight geometric elements and applies them to object movement.

Chapter 11, Manipulating the Light and Texture of Triangles, investigates the use of normal vectors in computer graphics and discovers how they are essential for representing, drawing, and lighting 3D models.

Chapter 12, Mastering Affine Transformations, reveals the primary set of operations that allows 3D points and models to be translated, scaled, and rotated in 3D space.

Chapter 13, Understanding the Importance of Matrices, discusses how transformations are represented as matrices and the power this brings to processing computer graphics.

Chapter 14, Working with Coordinate Spaces, teaches you about the different coordinate spaces a vertex is transformed through to get from its local coordinate system to a pixel on the screen.

Chapter 15, Navigating the View Space, guides you to explore ways to move the camera around in a 3D environment as well as discover some glitches in the matrix operations for rotations.

Chapter 16, Rotating with Quaternions, takes you deeper into the highly complex domain of quaternion rotations and shows how 4D spaces can solve 3D rotational issues.

Chapter 17, Vertex and Fragment Shading, guides you through the programming of shader code for the drawing of objects that transfers graphics from CPU control to be processed in parallel on the graphics card.

Chapter 18, Customizing the Render Pipeline, explores some fundamental techniques used to write shader code for elementary shaders that produce lighting effects.

Chapter 19, Rendering Visual Realism Like a Pro, reveals how light physically interacts with a 3D environment and the objects within it to develop a modern physically based rendering shader.

To get the most out of this book

The scripting language used in this book is Python. You should be familiar with coding in Python or at the very least have a working knowledge of an object-orientated programming language, such as C# or C++. Later in the book, when the OpenGL Shader Language is introduced, skills in this area are not required. Though an understanding of how procedural languages, such as C, are formatted would be beneficial. The projects created herein have been tested in the macOS and Windows environments.

Software covered in the book	Operating system requirements
Python	Windows or macOS
Pygame	Windows or macOS
PyOpenGL	Windows or macOS
PyCharm	Windows or macOS
OpenGL Shader Language (GLSL)	Windows or macOS

The downloading and setup instructions for this list of software are covered in the book as they are required.

If you are using the digital version of this book, we advise you to type the code yourself or access the code from the book's GitHub repository (a link is available in the next section). Doing so will help you avoid any potential errors related to the copying and pasting of code.

Download the example code files

You can download the example code files for this book from GitHub at `https://github.com/PacktPublishing/Mathematics-for-Game-Programming-and-Computer-Graphics`. If there's an update to the code, it will be updated in the GitHub repository.

We also have other code bundles from our rich catalog of books and videos available at `https://github.com/PacktPublishing/`. Check them out!

Download the color images

We also provide a PDF file that has color images of the screenshots and diagrams used in this book. You can download it here: `https://packt.link/rmsvT`.

Conventions used

There are a number of text conventions used throughout this book.

`Code in text`: Indicates code words in text, database table names, folder names, filenames, file extensions, pathnames, dummy URLs, user input, and Twitter handles. Here is an example: "The `color()` method takes the red, green, and blue channels, respectively, as arguments."

A block of code is set as follows:

```
import pygame

pygame.init()
screen_width = 800
screen_height = 200
screen = pygame.display.set_mode((screen_width,
                   screen_height))


done = False
white = pygame.Color(255, 255, 255)

pygame.font.init()
font = pygame.font.SysFont('Comic Sans MS', 120, False, True)
```

```
text = font.render('Penny de Byl', False, white)
while not done:
  for event in pygame.event.get():
    if event.type == pygame.QUIT:
      done = True
  screen.blit(text, (10, 10))
  pygame.display.update()
pygame.quit()
```

When we wish to draw your attention to a particular part of a code block, the relevant lines or items are set in bold:

```
background = pygame.image.load('images/background.png')
sprite = pygame.image.load('images/Bird-blue-icon.png')
while not done:
  for event in pygame.event.get():
    if event.type == pygame.QUIT:
      done = True
  screen.blit(background, (0, 0))
  screen.blit(sprite, (100, 100))
  pygame.display.update()
pygame.quit()
```

Any command-line input or output is written as follows:

```
tick=2, fps=714.2857055664062
tick=1, fps=714.2857055664062
tick=1, fps=714.2857055664062
tick=1, fps=666.6666870117188
tick=2, fps=666.6666870117188
```

Bold: Indicates a new term, an important word, or words that you see onscreen. For instance, words in menus or dialog boxes appear in **bold**. Here is an example: "Now, all you have to do is press **Play** to see the results."

> **Tips or important notes**
> Appear like this.

Get in touch

Feedback from our readers is always welcome.

General feedback: If you have questions about any aspect of this book, email us at `customercare@packtpub.com` and mention the book title in the subject of your message.

Errata: Although we have taken every care to ensure the accuracy of our content, mistakes do happen. If you have found a mistake in this book, we would be grateful if you would report this to us. Please visit www.packtpub.com/support/errata and fill in the form.

Piracy: If you come across any illegal copies of our works in any form on the internet, we would be grateful if you would provide us with the location address or website name. Please contact us at `copyright@packt.com` with a link to the material.

If you are interested in becoming an author: If there is a topic that you have expertise in and you are interested in either writing or contributing to a book, please visit `authors.packtpub.com`.

Download a Free PDF copy of this book

Thanks for purchasing this book!

Do you like to read on the go but are unable to carry your print books everywhere?

Is your eBook purchase not compatible with the device of your choice?

Don't worry, now with every Packt book you get a DRM-free PDF version of that book at no cost.

Read anywhere, any place, on any device. Search, copy, and paste code from your favorite technical books directly into your application.

The perks don't stop there, you can get exclusive access to discounts, newsletters, and great free content in your inbox daily

Follow these simple steps to get the benefits:

1. Scan the QR code or visit the link below

https://packt.link/free-ebook/9781801077330

2. Submit your proof of purchase
3. That's it! We'll send your free PDF and other benefits to your email directly

Part 1 – Essential Tools

This part will step you through the process of setting up and testing your development environment to follow along with the exercises throughout the book. Follow along step-by-step as we set up the development environment to be used throughout the book and test it by producing a graphics window of your choice in Windows or macOS. Following this, we will explore the mathematics required to draw common 2D shapes on the screen by coding the necessary formula into our project. As the part proceeds, more drawing functionality will be added until you have a 3D drawing environment with mouse and keyboard interaction that is capable of drawing, texturing, and lighting a 3D model.

In this part, we cover the following chapters:

- *Chapter 1, Hello Graphics Window: You're On Your Way*
- *Chapter 2, Let's Start Drawing*
- *Chapter 3, Line Plotting Pixel by Pixel*
- *Chapter 4, Graphics and Game Engine Components*
- *Chapter 5, Let's Light It Up!*
- *Chapter 6, Updating and Drawing the Graphics Environment*
- *Chapter 7, Interactions with the Keyboard and Mouse for Dynamic Graphics Programs*

1
Hello Graphics Window: You're On Your Way

Students new to games and computer graphics fall into two camps: ones that think mathematics is a real bore and of no real use, and others that realize it is akin to breathing. The fact is that if you want to be successful (and stay alive) in the domain, then you can't deny the importance of mathematics. This doesn't mean that learning about it should be a grind. What we find most exciting about teaching and learning about mathematics in the computer games and graphics space is that the equations and numbers come alive through visual representations that remain otherwise unseen in other fields. Throughout this book, you will learn the essential mathematics in games and graphics by exploring the theory surrounding each topic and then utilizing it in real-world applications.

To this end, throughout this book, you will not only gain an understanding of the essential mathematics that is used throughout games and graphics, but also apply its principles in one of today's hottest programming languages: **Python**.

> **Note**
> As the content of this book relates to both computer graphics and computer games, rather than continually typing out the laborious phrase "computer graphics and computer games," we will endeavor to refer to both collectively as *graphics*.

"Why Python?" we hear you ask. *"Why not some fancy game's engine?"* Well, besides having Python programming as a great skill under your belt when it comes to applying for a job, Python with the assistance of **PyCharm** (our **Integrated Development Environment – IDE**), **Pygame** (a graphics interface), and **PyOpenGL** (a Python/OpenGL API) reveals much of the underlying technical processes hidden by game engines and brings you face to face with direct calls to the graphics APIs that access OpenGL and DirectX. It's a more challenging way to learn about mathematics if you remove all the high-level methods available to you that hide how your applied mathematics programming skills are working.

Learning about the mathematics of computer games and graphics at this level is essential in your learning journey. In this ever-changing domain, as technology progresses, your theoretical level of the *how's* and *why's* of the topic can remain constant and make you a highly skilled and adaptive programmer. A deep-seated knowledge of the algorithms and equations running the APIs of contemporary games and graphics engines is critical and transferable knowledge that developers can take from one platform to another and from one API to another. It not only provides extra insight while troubleshooting and debugging, but also gives a programmer an intuitive understanding of what functionality an API might contain without actually knowing that API. It's akin to understanding all the tools in Adobe Photoshop and being able to perceive what tools will be available in Affinity Photo without ever using it. Having good math skills blurs the line between what is possible and what is not.

Now that you know why you are learning about mathematics in the way presented throughout this book, it's time to dive into getting your own development environment set up and exploring the anatomy of a simple graphics window that will become the basis of all our exercises moving forward. We will begin with a guided exercise in getting your own development environment setup using PyCharm. This will allow you to jump right into coding with as little fuss as possible. It removes the need to manually download and install packages and plugins that usually require knowledge of command-line install procedures, which can become laborious and vary between desktop machine setups. This means more time spent on coding to start using Python to open windows and explore the range of fundamental graphics drawing methods that will set you up for graphics development success.

In this chapter, we will cover the following topics:

- Getting Started with Python, PyCharm, and Pygame
- Creating a Basic Graphics Window
- Working with Window and Cartesian Coordinates

Technical requirements

As mentioned in the introduction, we will be using the programming language, Python, along with an editor called PyCharm and a plugin/graphics library called Pygame. You should be a programmer to embark on the content of this course but not necessarily a Python programmer, as someone with working knowledge in any procedural language will pick up Python quickly enough.

These are all cross-platform tools and can be used on Windows, macOS, and Linux. We'll leave it up to you to find the relevant versions of the software for your machine, however, to get you started, download Python from python.org. As shown in *Figure 1.1*, under the **Downloads** tab, you can get the latest version of the software. At the time of writing, it is Python 3.10.0. It is highly recommended that you follow along with the exercises in this book using the versions shown in the book to reduce any errors you may come across in replicating the code. If you are reading this and there is a later version of Python available, you can find version 3.10.0 under the **All releases** option as follows:

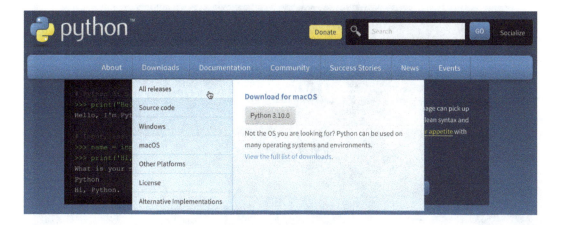

Figure 1.1: The Downloads tab at python.org

Thus, the steps to get Python installed on your machine are as follows:

1. Visit `python.org` and download a version of Python relevant to your operating system.

2. Run the downloaded file and follow the directions to install the software.

In order to code and compile with Python, we require an editor; for this, we will be using PyCharm. The same advice stands for version numbers of this software. At the time of writing this book, PyCharm is on version 2021.3. There is a *Profession* and a free *Community* version of PyCharm, but we will only require the free version for this book. To install PyCharm on your machine, do as follows:

1. Visit `www.jetbrains.com/PyCharm/download`.

2. Download a version of the installer relevant to your machine.

3. Run the installer to set up PyCharm.

Now that you have the fundamental software installed, we will proceed to create a basic graphics window before learning how to draw on the screen.

Getting Started with Python, PyCharm, and Pygame

In this section, we will go through the process of setting up PyCharm and rendering a graphics window on the screen. To do this, follow these steps:

1. Open PyCharm. The first window will appear as shown in *Figure 1.2*.

Figure 1.2: The opening PyCharm screen on an Apple Mac (the screen
on other platforms will look similar but not the same)

You may want to take some time to look at the **Customize** section. This will allow you to set up colors, fonts, and other visual components. There are also many different settings and customizations you can make within PyCharm, and if you ever want to explore these, then we recommend a visit to the manual at `www.jetbrains.com/help/PyCharm/quick-start-guide.html`.

2.   Next, select the **New Project** button on the **Projects** page of the startup window. A window like what is shown in *Figure 1.3* will appear.

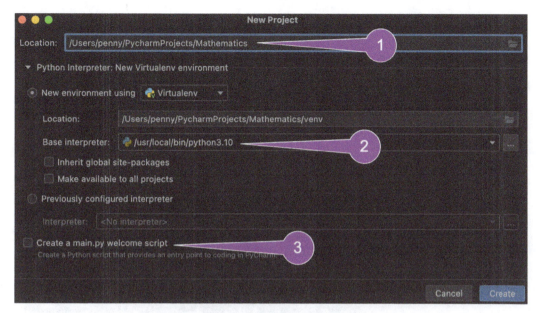

Figure 1.3: PyCharm's New Project window and settings

3. Next, we need to set up our new project as follows:

   I.   For **Location** (**1**), create a new folder for your files. Unlike a single source file such as what you might get with a Word document, a Python project can consist of many files. Therefore, instead of specifying a single file name, you must specify a folder. Call it *Mathematics*.

   II.   In the field for **Base interpreter** (**2**), ensure that you select the version of Python that you downloaded and installed. It's possible to have more than one version of Python installed on your machine. This is useful if you are working on different projects for clients as you can set the version when you create a new project.

   III. When you first open the **New Project** window, the **Create a main.py welcome script** tickbox (**3**) will be ticked. We don't require a default main.py script and therefore you should untick it.

4. Once the project settings have been entered, click on the **Create** button.

5. As shown in *Figure 1.4*, after PyCharm opens the new project window, it will display a column on the left with the folder structure (**1**). By clicking on **Python Packages** at the bottom of the window (**2**), a new section will be revealed where you can search and install packages. In the search section (**3**), type *Pygame*. This will bring up a list of packages below the search area. Select the one at the top that says *Pygame*, and then on the right, click the **Install** button (**4**). Once Pygame has been installed, you can close the packages section (**5**).

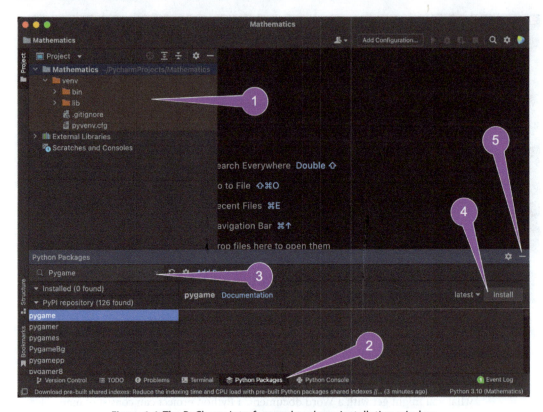

Figure 1.4: The PyCharm interface and package installation window

Python, PyCharm, and Pygame are now set up, and you are ready to create your first graphics window.

> **For those new to Python**
>
> If you are new to Python but not programming, it is a straightforward language to pick up. To get up to speed quickly, any newbies are encouraged to read over the beginner's documentation at `https://wiki.python.org/moin/BeginnersGuide`.

Creating a Basic Graphics Window

It's time to create your first basic graphics window in PyCharm. To do this, perform the following steps:

1. Right-click on the project folder (**1**), as shown in *Figure 1.5*.

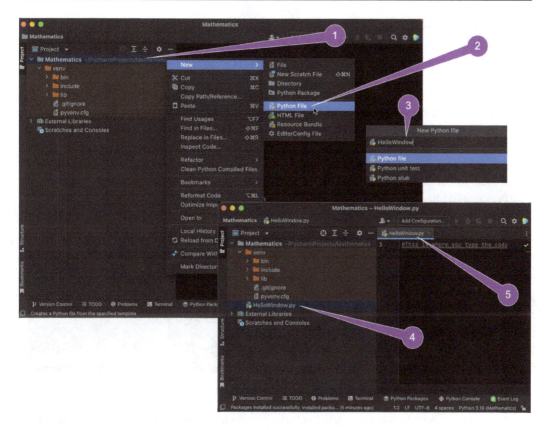

Figure 1.5: Creating a new Python script file (steps 1-5)

2. Select **New** > **Python File** *(2)*.

3. In the popup window, enter the name of your new Python script (**3**). In this case, it will be `HelloWindow.py`. There's no need to add the `.py` suffix as PyCharm will add this for you. Hit the *Enter/Return* key.

4. Once the file has been created, you will see it in the **Project** *section* of PyCharm (**4**).

5. The file will also be open and displayed on the left as a tab title (**5**). The window below the tab is where you will type in your code as follows:

```
import pygame
pygame.init()
screen_width = 1000
screen_height = 800
screen = pygame.display.set_mode((screen_width,
                screen_height))
```

The preceding code imports the Pygame library, so you can access the methods provided by Pygame to set up graphics windows and display images. The specific method used in this case is set_mode(), which creates and returns a display surface object.

> **Tool tip**
>
> There are no semi-colons at the end of lines in Python. Although, if you accidentally use one because it's second nature to you as a programmer, then it's very forgiving and ignores them.
>
> The editor will also insist on variables and methods being written in snake_case (www.codingem.com/what-is-snake-case/) where separate works are joined by a delimiter, such as an underscore. You can see this format in the previous code in the variable named screen_width.

When complete, the window will look like *Figure 1.6*.

```
HelloWindow.py
1   import pygame
2   pygame.init()
3   screen_width = 1000
4   screen_height = 800
5   screen = pygame.display.set_mode((screen_width, screen_height))
```

Figure 1.6: Code entered into PyCharm to generate a basic graphics window

It is not the aim of this book to teach you Python or the ins and outs of the Pygame library as we want to focus on mathematics; however, short explanations will be given as we add new features. In this case, the first line ensures that the functionality of the Pygame package is included with the code, which allows us to use the methods embedded inside Pygame. Next, we initialize the Pygame environment on line 2. Finally, we create a screen (a graphics window or surface also known as a display). The width of this window is 1,000 pixels by 800 pixels in height.

If you are unsure of what the parameters of a method are, simply mouse over/hover your cursor the method and PyCharm will pop open a hint box with more information about the method.

6. The code can now be run. To do this, right-click on the HelloWindow.py filename in the **Project** window and then select **Run 'HelloWindow'**, as shown in *Figure 1.7*:

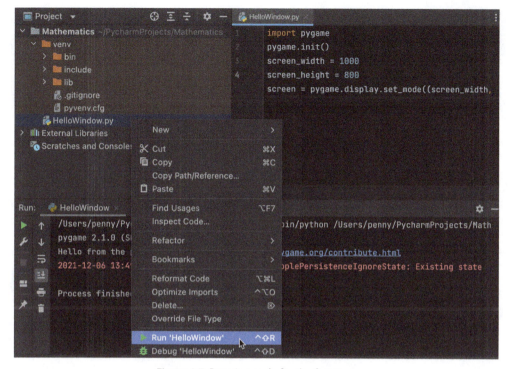

Figure 1.7: Running code for the first time

If you've typed in the code correctly, a window with dimensions of 1,000 x 800 will pop open and then immediately close. It might be too fast to see, so try running it again. You will know the program has been completed without error as the console now visible at the bottom of the PyCharm window will display an exit code of 0, as shown in *Figure 1.8*:

Figure 1.8: When a program runs without error, it will return an exit code of 0

Note that once you've run code for the first time, it becomes available as a run option in the top toolbar and can be easily run again using the green play button at the top of the PyCharm window. Ensure that the filename next to the button is the one you want to run, as shown in *Figure 1.9*:

Figure 1.9: Running the previous code again

7. In its current state, the graphics window isn't much help to us if it won't stay open. Therefore, we need to force it to stay open. You see that as the code executes line by line, once it reaches the end, the compiler considers the program to be finished. All graphics windows need to be forced to stay open with an endless loop that runs until a specific condition is reached. Game engines inherently have this built in and call it the main game loop. As you will discover in *Chapter 6, Updating and Drawing the Graphics Environment*, the main loop is the heartbeat of all graphics programs. In this case, we must create our own. To do this, modify the code by adding the highlighted text to your program as follows:

```python
import pygame
pygame.init()
screen_width = 1000
screen_height = 800
screen = pygame.display.set_mode((screen_width,
                   screen_height))
done = False
while not done:
    for event in pygame.event.get():
        if event.type == pygame.QUIT:
            done = True
    pygame.display.update()
pygame.quit()
```

Tool tip

Instead of relying on braces, for example, '{' and '}' to segment code blocks, Python relies on indentation, and at the end of a line with a Boolean expression, you will always find a full colon (:).

You've just created an endless loop that will keep running until the window is manually closed by the user. The `done` variable is initially set to false, and only when it becomes true can the loop end. `pygame.event.get()` returns a list of events the window has received since the last loop. These can be keystrokes, mouse movements, or any other user input. In this case, we are looking for `pygame.QUIT` to have occurred. This event is signaled by the attempted closing of the window. The final `pygame.quit()` method safely closes down all Pygame processes before the window closes.

8. Run the program again. This time, the graphics window will stay open. It can be closed using the regular window closing icons that appear on the top bar of the window (on the left in Mac OS and on the right in Windows).

9. Because code will be reused and updated from chapter to chapter, the best way to organize it all would be to place all the code for a chapter into its own folder. So, let's create a folder for `Chapter One`. To do this, right-click on the **Mathematics** project folder at the top and then select **New > Directory**, as shown in *Figure 1.10*. Name the folder **Chapter_One**.

10. Drag and drop `HelloWindow.py` into the new folder. PyCharm will ask you to refactor the code and it's fine to go ahead and do so. You can now keep all the code from any chapter in one place even when the same script names are used.

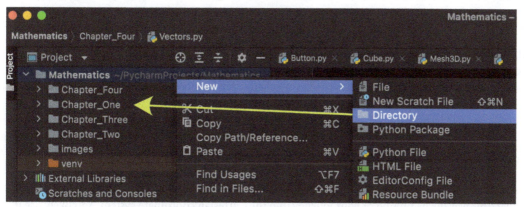

Figure 1.10: Creating a new directory inside PyCharm

And that's it. Your first basic graphics window. Keep a copy of this code handy as you will need to use it as the basis of all projects from this point forward. Whenever we ask you to create a new Python file in PyCharm, this code will be the jumping-off point.

Hints

To further help you with creating a window, consider the following points:

- If you haven't used Python before but are familiar with other programming languages, the absence of brackets and braces for the sectioning of code will feel a little odd at first. Instead of brackets, Python uses indentation. You can read more about this at `www.geeksforgeeks.org/indentation-in-python/`.

- Python does not need semicolons at the end of lines.

- Whenever you encounter methods or properties used in the code that you are unfamiliar with, it is a good exercise for your own independent learning to investigate these in the Python and Pygame API listings as follows:

- For Pygame, see `www.pygame.org/docs/genindex.html`.

- For Python, see `docs.python.org/3/library/index.html`.

- As you may have noticed as you were typing out the code, Python can be fussier than other programming languages when it comes to formatting and syntax. PyCharm will give you suggestions for most formatting issues by showing warning symbols at the end of problem lines or drawing a rippled line under methods and variables. A right-click on these will reveal suggestions for fixes. For a more extensive list, however, see `www.python.org/dev/peps/pep-0008/`.

This section has been a whirlwind of an introduction to opening a basic graphics window. Knowing how to achieve this is fundamental to the exercises in the remainder of this book or the start of your own graphics or game projects, as it is the key mechanic that supports all these types of applications. In the next section, we will expand on your knowledge of the graphics window by exploring its coordinates and layout.

Working with Window and Cartesian Coordinates

Remember drawing an *x*- and *y*-axis on paper in school and then plotting a point based on its x and y value? This was the beginning of your journey into the Cartesian coordinate system. You likely use it every day to locate positions on maps, in drawings, or on your computer screen. Fundamentally, it is a way of uniquely representing a location with a single value on a line, a pair of numerical coordinates on a plane (2D), or with a triplet of values in 3D. In fact, Cartesian coordinates even extend into four, five, and upward dimensions, though they aren't relevant here.

The axes of the system have evenly spaced measurements, much like a ruler, and are used to determine where a location made up of coordinates is placed. For example, in one dimension there is only one axis, and going with popular convention, we will call this x. To plot x = 5 is to locate the value of 5 on the axis and place a point there. In two dimensions with the axes, x and y, and the coordinate (2, 3) where x is the first value and y is the latter, a point is placed where these values intersect. These examples are illustrated in *Figure 1.11*.

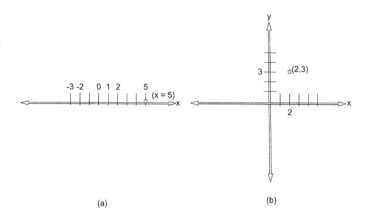

(a)

(b)

Figure 1.11: Cartesian plots of x = 5 in one dimension (a) and (2,3) in two dimensions (b)

The majority (if not all) of graphics applications work in two and three dimensions and as such we will reduce our discussion in this book to those with most applications in 3D. While the computer screen is 2D, it is also used to fake 3D; however, the mathematics applied still occurs within 3D calculations with the *x*-, *y*-, and *z*- axes. In working with graphics, we must therefore be aware of different ways of viewing Cartesian coordinates. In the next exercise, we will show the plotting of points in 2D and how Cartesian coordinates work in graphics.

Let's do it...

To create a window with some plotted points defined by Cartesian coordinates, we perform the following steps:

1. Create a new Python script called `PlotPixel.py` and add the following code (the differences between `HelloWindow.py` and this script are shown in bold):

```
import pygame
pygame.init()
screen_width = 1000
screen_height = 800
screen = pygame.display.set_mode((screen_width,
               screen_height))
done = False
white = pygame.Color(255,255,255)
while not done:
   for event in pygame.event.get():
     if event.type == pygame.QUIT:
```

```
        done = True
    screen.set_at((100, 100),white)
    pygame.display.update()
  pygame.quit()
```

2. The `set_at()` method will plot a single pixel point in the color of your choosing; in this case, the pixel will be white at the coordinates x = 100, and y = 100. To see this in action, run the script. Remember that you will need to right-click on the filename in the **Project** window and select **Run** before you'll be able to run it from the little green icon at the top-right of the window. Now, try plotting another point at (200, 200). You can do this with a second `set_at()` call, for example, as follows:

```
screen.set_at((200, 200),white)
```

The resulting screen image is shown in *Figure 1.12* (note that the pixel locations have been enhanced with white squares so you can see them in print).

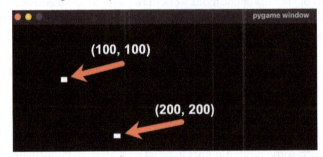

Figure 1.12: Pixel locations when plotted in a default window

What did you notice about the location of the pixels? Based on their position, where would you conclude the origin (0,0) of the window to be?

If you said the upper-left corner of the window, you would be correct. We all learn in school that the positive Cartesian coordinate system runs for positive x values to the right along the page and for positive y values up the page. However, in computer graphics, this isn't true, and the coordinate systems will follow different conventions depending on the platform and tools. In this case, the x coordinates run as you would expect but the y values are inverted with positive y values going down the page.

Why? Because early graphics displays used physical, electron scan lines beginning in the upper-left corner of the display, scanning across, and then down. It doesn't need to be like this anymore, but it just is.

Most game engines transform these coordinates so that the y-axis faces up.

However, as the astute computer graphics programmer that you are, it's always best to figure out which way is up according to the coordinate system you are working with, either through the documentation or by running a few experiments of your own.

3. If you would prefer the origin of the coordinate system to be shown in the lower left-hand corner, then you can modify the coordinates with a simple method as follows:

```python
import pygame


pygame.init()
screen_width = 1000
screen_height = 800
screen = pygame.display.set_mode((screen_width,
                 screen_height))
done = False
white = pygame.Color(255, 255, 255)



def to_pygame_coordinates (display, x, y):
    return x, display.get_height() - y



while not done:
    for event in pygame.event.get():
        if event.type == pygame.QUIT:
            done = True
                screen.set_at(to_pygame_coordinates
                            (screen, 100,100),
                            white)
                screen.set_at(to_pygame_coordinates
                            (screen, 200,200),
                            white)
    pygame.display.update()
pygame.quit()
```

In the preceding code, the new `to_pygame_coordinates()` method flips the y coordinate based on the screen height and returns a new set of (x, y) values for plotting the pixel basing the origin of the screen in the lower-left corner of the window. Though, after you've worked with graphics for a while, you'll get used to the upper-left origin and as such we will assume the origin of the coordinates (0,0) to be the upper-left of the graphics window, unless stated otherwise.

Drawing rectangles

In the preceding figure, we used rectangles to enhance the size of the pixels to make them visible. To draw a point in a location in the window that is larger than a single pixel, we can draw a small rectangle or square instead. To do this, the code is as follows:

```
import pygame

pygame.init()
screen_width = 1000
screen_height = 800
screen = pygame.display.set_mode((screen_width, screen_height))
done = False
white = pygame.Color(255, 255, 255)


def to_pygame_coordinates(display, x, y):
  return x, display.get_height() - y



while not done:
  for event in pygame.event.get():
    if event.type == pygame.QUIT:
      done = True
        position = to_pygame_coordinates
                      (screen, 100, 100)
        pygame.draw.rect(screen, white,
                      (position[0],position[1],
                      10, 10))
  pygame.display.update()
pygame.quit()
```

The definition of pygame.draw.rect can be found at https://www.pygame.org/docs/ref/draw.html#pygame.draw.rect. In short, a rectangle requires a screen, color, a starting location (identified by position in the code), and a width and height (we are using 10 in this example).

This task has shown you how to work with a window and the coordinates used within it. It's a valuable skill as pixel-by-pixel is how every graphics application is written. It's the nature of the display devices we use.

Your Turn...

> **Note**
>
> After most code-along exercises in the *Let's do it...* sections of this book, you will find a section like this that encourages you to apply the skills you've learned thus far. The answers to these exercises can be found at the end of the chapter.

Exercise A.

Find an image of the constellation representing the Leo star sign and draw those series of stars on the screen. An example of its layout is shown in *Figure 1.13*. Notice that we have also included a size factor for each star based on its relative brightness, where we used rectangles of different sizes to draw each one.

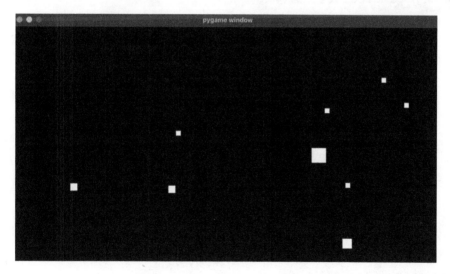

Figure 1.13: The constellation of Leo

When you have completed this and reviewed the answer at the end of the chapter, why not try and recreate your own star sign?

Summary

You are now an expert at opening graphics windows and have new-found skills at plotting pixels in that window using Cartesian coordinates. This skill is the essential underpinning of everything that gets drawn on a computer screen; it's through the understanding of mathematics that you will gain throughout this book that you will be able to expand that single pixel into a world of complex 3D structures, colors, and animations. Some of these basic graphics primitives will be explored in the following chapter.

Answers

Exercise A:

```
import pygame

pygame.init()
screen_width = 900
screen_height = 500
screen = pygame.display.set_mode((screen_width, screen_height))
done = False
white = pygame.Color(255, 255, 255)


def draw_star(x, y, size):
  pygame.draw.rect(screen, white, (x, y, size, size))


while not done:
  for event in pygame.event.get():
    if event.type == pygame.QUIT:
      done = True

  draw_star (121, 320, 15)
  draw_star (327, 324, 15)
  draw_star (691, 431, 20)
  draw_star (697, 317, 10)
  draw_star (626, 246, 30)
  draw_star (343, 212, 10)
  draw_star (653, 165, 10)
  draw_star (773, 102, 10)
  draw_star (822, 153, 10)
  pygame.display.update()
pygame.quit()
```

2
Let's Start Drawing

There's nothing quite like getting your first graphics application running. There's a real sense of achievement when you see your ideas for the rearrangements of pixels come together on the screen. My first coding language was BASIC on an Amstrad CPC664 and even though the code was quite laborious, I liked nothing better than to draw shapes and change the colors on the screen.

Computer-drawn images ultimately end up as single pixels (such as the ones drawn in the previous section) with differing colors on a graphics display. To draw all objects pixel by pixel whenever you'd like to create an image would be a long, drawn-out process. It might have been how images were rendered on a screen in the 1950s, but now, with advanced technology, Cartesian coordinate systems, and mathematics, we are able to specify drawing primitives and use these over and over to compose a picture.

In this chapter, you will learn about the most primitive of concepts and shapes upon which all graphics are built and discover how to work with them in Python and Pygame. The overall goal will be to provide you with an understanding of, and practical skills in, the application of these concepts, which will form the basis for everything you create moving forward.

In this chapter, you will learn about these fundamental concepts through the exploration of the following:

- Color
- Lines
- Text
- Polygons
- Raster images

Technical requirements

The exercises in this chapter use the same Python and PyCharm software and configurations as were set up in *Chapter 1, Hello Graphics Window: You're On Your Way*. Remember to keep the code from each chapter separate. Before you continue reading, be sure to make a new folder to place the code in for this chapter.

The solution files containing the code can be found on GitHub at `https://github.com/PacktPublishing/Mathematics-for-Game-Programming-and-Computer-Graphics/tree/main/Chapter02`.

Color

Computer graphics, by their very definition, are a visual medium and as such rely on color to be seen. A **color space** is the range of colors that can be generated from the primary colors of a device. For example, imagine you are creating an oil painting. Say you had a tube of red, a tube of yellow, and a tube of blue, and you mixed them together in every conceivable ratio, then all the resulting colors would be the color space for that set of three tubes. If your friend did the same thing with slightly different reds, yellows, and blues, they would produce another color set unique to them. If you both mixed a green from equal parts of your own blue and yellow, you would get green that was 50% blue and 50% yellow, but because you both started with slightly different versions of blue and yellow, the resulting greens would be different.

The same applies to computer screens, printers, cameras, and anything else with a colored screen or print. The color you consider green on one monitor might look completely different on another and it may look different again when you print it on paper. The range of colors that can be created by mixing paints or seen on a computer screen is called a **gamut**. Given the previous example with the tubes of paint, we could say that you and your friend are working with different color gamuts.

To help alleviate this problem, there exist numerous standard color spaces and they dictate how a color should appear. The most common one is **RGB**, whose initials stand for **Red**, **Green**, and **Blue**. All colors displayed on a computer screen can be defined by mixing different amounts of red, green, and blue.

The values of each color are specified in one of two ranges, either 0 to 1 or 0 to 255. For values between 0 and 1, a 0 indicates none of the color and a 1 all the color. Given that the RGB value is specified as a tuple in the form (red, green, blue), (0.5, 0, 1) would denote a color with a half strength of red, no green, and a full strength of blue. In a similar vein, where the color values are specified in the range 0 to 255, the same color would be represented by (128, 0, 255), where you have half the amount of available red mixed with the full amount of blue.

It's most likely you've experienced mixing RGB colors before, as you can set them in many software packages for anything from text to paintbrushes. *Figure 2.1* shows the color picking tools in Adobe Photoshop and Microsoft Word and points out where the RGB values are set:

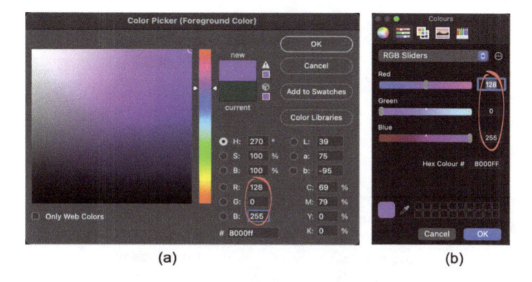

Figure 2.1: Color pickers from popular software: (a) Adobe Photoshop and (b) Microsoft Word

---

Hint

When setting a color while coding, if you aren't sure what the RGB values are for the color you want, use a color picker from any software you have available to you. If the coding language you are using takes color values between 0 and 1 instead of between 0 and 255, simply divide the values by 255. For example, a color specified as (100, 50, 90) for color values between 0 and 255 would become (100/255, 50/255, 90/255) = (0.39, 0.20, 0.35).

Let's do it...

In this exercise, we will investigate the use of color for displaying a rainbow of pixels across the screen:

1. Create a new Python file called RBGSpace.py and add the basic starter code created in *Chapter 1, Hello Graphics Window: You're On Your Way*, to open a window.

2. Pygame takes color values in the range 0 to 255. Although you can't easily display the blending of three colors in a two-dimensional window, you can still explore the range between two of the color values with a program like this where the code you need to add the basic starter code is displayed in bold:

```
import pygame

pygame.init()
screen_width = 1000
```

```
screen_height = 800
screen = pygame.display.set_mode((screen_width,
                   screen_height))
done = False

while not done:
  for event in pygame.event.get():
    if event.type == pygame.QUIT:
      done = True
  for y in range(800):
    for x in range(1000):
      screen.set_at((x, y),
          pygame.Color(int(x/screen_width * 255),
            int(y/screen_height * 255), 255))

    pygame.display.update()
  pygame.quit()
```

The `color()` method takes the red, green, and blue channels respectively as arguments. You will notice that the red and green channel values will vary while the blue channel stays fully turned on with the value 255.

The result shown in *Figure 2.2* shows you the range of colors by mixing all the reds with all the greens, while the blue channel is turned on to the maximum:

Figure 2.2: The range of colors available by mixing across the
red and green color channels with blue fully on

Your turn…

Exercise A: Draw the range of colors that mix the green and blue channels with no red.

As discussed, graphics would be nothing without color. In the previous exercise, we used it for coloring pixels. However, there are many other graphics constructs that build upon the simple pixel and so let's now move on to discover another graphics fundamental: lines.

Lines

The drawing of lines holds a very special place in my heart on my computer graphics learning journey. While you were back in school learning about **Cartesian coordinates**, you most likely also learned how to draw a straight line and that the relationship of the y coordinate to the x coordinate is as follows:

$$y = mx + c$$

Where m is the **gradient** (or **slope**) and c is the **y-intercept**. The gradient refers to how vertical or horizontal the line appears. A perfectly horizontal line has a gradient of 0, whereas a perfectly vertical line has a gradient of infinity. The y-intercept is the value on the y axis (where x = 0) that the line crosses.

> **Note**
>
> For a refresher on the ways to represent a line mathematically, see `https://en.wikipedia.org/wiki/Linear_equation`.

Exploring line equations makes more sense when you can see them on the screen and investigate the effect of modifying their variables. Lucky for us, there is freely available software on the web that can help with this.

Let's do it…

We can visualize this relationship of the variables in a line equation in DESMOS with the following steps:

1. Visit `https://www.desmos.com/calculator`.
2. In the equation box shown in *Figure 2.3*, type `y = 2x + 4` to see the same graph of the line plotted:

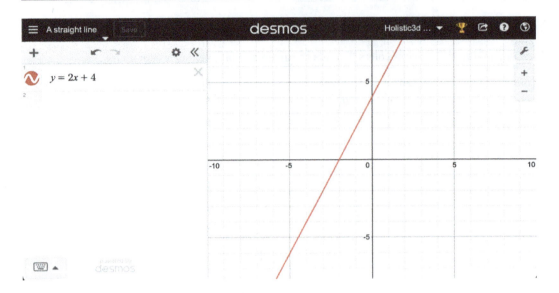

Figure 2.3: Plotting a straight line from its equation in DESMOS

The value of 2 in the equation is the slope (this is how steep the line is). The smaller the value, the flatter (more horizontal) the line.

3. To try making the slope more horizontal, change the value of 2 to 1. A value of 1 gives you a 45-degree line. You can go even lower if you like, with 0.5, or if you use 0, you will get a perfectly horizontal line.

4. The value of 4, which is the **y-intercept** in this case, in *Figure 1.14* adjusts where the line crosses the y axis. Change this as you see fit to witness the same changes in the line.

After this short introduction to DESMOS, you'll now have the skills to use the software for evaluating other equations including lines. This will come in handy as we progress through the book, and you might need to check whether what you are drawing on the screen is correct. For example, if you believe your code should be drawing a horizontal line and it is drawing one at 45 degrees, then you can use tools such as DESMOS to double-check your working out.

Your turn...

Exercise B: Use DESMOS to plot a line with a slope of 8 and a y-intercept of 0.5. Now that you are aware of the basic anatomy of a line, let's have a look at the way to plot them with Pygame. Unlike DESMOS, which takes the equation of a line, Pygame requires just two points: the starting point and the end point. It takes a lot of the heavy lifting out of drawing a line, as you will discover in the section *Line Plotting Pixel by Pixel*.

Let's do it...

You are about to discover just how easy it is to plot a line with Pygame. In this exercise, we will also include mouse clicks to add a dynamic factor to line drawing by allowing a user to select the line endpoints:

1. Create a new Python file called `PygameLine.py`.

2. Add the following code, which uses the built -in Pygame line method:

```python
import pygame
from pygame.locals import * #support for getting mouse
                            #buttons

pygame.init()
screen_width = 1000
screen_height = 800
screen = pygame.display.set_mode((screen_width,
                                  screen_height))

done = False
white = pygame.Color(255, 255, 255)
times_clicked = 0

while not done:
  for event in pygame.event.get():
    if event.type == pygame.QUIT:
      done = True
    elif event.type == MOUSEBUTTONDOWN:
      if times_clicked == 0:
        point1 = pygame.mouse.get_pos()
      else:
        point2 = pygame.mouse.get_pos()
      times_clicked += 1
      if times_clicked > 1:
        pygame.draw.line(screen, white, point1,
                         point2, 1)
        times_clicked = 0
  pygame.display.update()
pygame.quit()
```

> **Tool Tip**
>
> A # is used in Python to begin a comment.

This code uses a variable called `times_clicked` to keep track of how many times the mouse button has been clicked. These clicks trigger the MOUSEBUTTONDOWN event. Each time the mouse is depressed, a point is recorded. After two clicks, the two points collected are used as the opposite ends of a line, which in this case is drawn in `white` with a thickness of 1 as shown in *Figure 2.4*.

> **Note**
>
> For more information on the pygame.draw library, see the Pygame documentation at https://www.pygame.org/docs/ref/draw.html.

As you've discovered, Pygame makes it easy to draw a colored line between two points:

Figure 2.4: A straight line produced by Pygame

It also makes it easy to create more graphical text effects, as you are about to discover.

Text

It's rare that you will find a graphical user interface without text. In the past, display devices operated in two modes, a text mode, and a graphics mode. If you ever used the original versions of DOS or Unix, you were operating in text mode. This meant there was no way to draw graphics on the screen (other than the characters of the text themselves), so text couldn't be written on buttons or appear upside-down or sideways and the design of the characters themselves was limited. The fanciest images available on a screen were drawn with text, such as this fish:

```
|\ \\\\__   o

|\_/  o\  o

>_  (( <_  oo

| / \__+___/

|/   |/
```

The images, called ASCII art (see www.asciiart.eu for more examples), originated in the 1970s and 1980s on computer bulletin board systems, but their popularity waned as computer displays and graphics cards became more advanced and graphical interfaces were introduced. However, people still use ASCII art in text-based messaging situations such as chat in multiplayer online games, email, and message boards.

Nowadays, text is drawn on the screen, meaning that it is graphics based and as such we have access to a variety of attributes that we can change to define how it looks. Some of these include the typeface, color, size, spacing, and orientation.

The **typeface**, more commonly referred to as a **font**, is a specific set of characters that share the same style. Some of the more commonly known ones are Arial, Verdana, and Times New Roman. There are two categories font design falls into: **serif** and **sans serif**. A serif is a decorative line that embellishes the font, with small lines on the ends of character strokes, as seen in *Figure 2.5*:

Times New Roman
is a serif font.

Arial
is a sans-serif font.

Figure 2.5: A serif font and sans serif font

For graphical displays, while it's quite acceptable and possible to use serif fonts, sans-serif fonts are easier to read on computer displays.

The following code demonstrates how to draw text with Pygame (with the result shown in *Figure 2.5*):

```python
import pygame

pygame.init()
screen_width = 800
screen_height = 200
screen = pygame.display.set_mode((screen_width,
                                  screen_height))

done = False
white = pygame.Color(255, 255, 255)

pygame.font.init()
font = pygame.font.SysFont('Comic Sans MS', 120, False,
                           True)
text = font.render('Penny de Byl', False, white)
while not done:
  for event in pygame.event.get():
    if event.type == pygame.QUIT:
      done = True
  screen.blit(text, (10, 10))
  pygame.display.update()
pygame.quit()
```

First, pygame.font.init() is used to initialize the use of fonts, which only needs to be called once per program. Next, the typeface is set up; in this case, it is Comic Sans MS.

For a list of all the system fonts printed to the console you can use this:

```python
print(pygame.font.get_fonts())
```

Next, font.render creates a display of text that uses the font. The parameters following the string are a bool for antialiasing and the color. At this point, the text isn't displayed but only set up.

Inside the while loop, you will find a call to screen.blit. This places the text on the screen at the given (x, y) location.

You might be wondering why text needs a `screen.blit` when drawing a line doesn't. The purpose of `screen.blit` is to draw an existing drawn surface on top of an existing one. You can think of a drawn surface in this case as an image with transparency. The text is drawn into an image called text with the `font.render` method. So, it's being held in memory. `screen.blit` copies text on to the display window pixel by pixel, combining the colors in text with what is already on the screen. We will see further use of `screen.blit` when we start loading in external images.

The result of our example is shown in *Figure 2.6*:

Figure 2.6: Pygame being used to draw text

To use a custom font with Pygame, first you need a font definition file. The file format for these is **TrueType font** and they have a suffix of `.ttf`. It's possible to create your own with software, however, for the point of the next exercise, you can download a free one.

Let's do it...

In this exercise, you will learn how to take a downloaded font and use it with Pygame to display text in the graphics window:

1. Visit www.dafont.com, select a font, and download the `.ttf` file.

2. Drag and drop the font file (while holding the *Ctrl* key) into PyCharm's project window. The software might ask you to perform a refactor – just say yes.

3. Create a new Python file called `CustomText.py` and start with the final code from the previous exercise.

4. Replace the line beginning `font  =` with the following:

    ```
    font = pygame.font.Font('fonts/Authentic Sheldon.ttf',
                            120)
    ```

    Here, the string is the name of your font file. Note that it is possible to create a new folder inside the project window and place the font file in there if you'd like to keep things organized. To get the path to the font file, wherever it is located, right-click on the font filename and copy its path from the content root as shown in *Figure 2.7*:

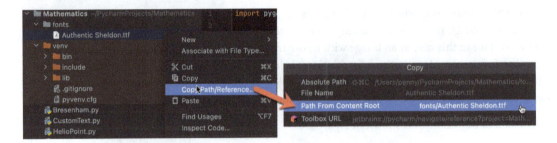

Figure 2.7: How to get the path and filename of a file in PyCharm

5. Now all you have to do is press **Play** to see the results, as illustrated in *Figure 2.8*:

Figure 2.8: Using a TrueType font to change the font face with Pygame

Now that you know how, placing text into a window using any font will become common place. In graphics applications, text is most widely used for graphical user interfaces. The next challenge will help reinforce your understanding of font use.

Your turn...

Exercise C: Find another **TrueType font** file that you like. Use it to display another line of text below the one already coded.

> **Note**
> If you are interested in creating your own font from scratch, take a look at the online font editor from `fontstruct.com`.

As you might imagine, the mathematics involved in drawing a font are quite complex. They can be drawn using a series of curves or, more specifically, **Bézier curves**, which we will examine in *Part Three: Essential Geometry*. But for now, we are containing our discussion of graphics elements to those constructed of straight lines, and the polygons we will look at in the next section are a perfect example.

Polygons

A polygon is a flat shape defined by several straight lines connected to enclose an area. The polygon with the least number of sides is a triangle. A generic polygon is called an **n-gon** where *n* is the number of sides. A triangle is a 3-gon, whereas squares and rectangles are 4-gons. The side of a polygon is a straight line and is called an edge. The points at which they connect are called vertices.

Polygons are classified according to their shapes, as shown in *Figure 2.9*:

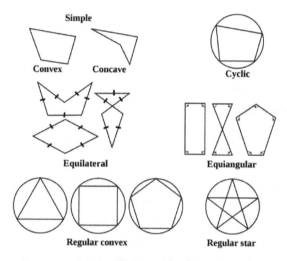

Figure 2.9: Types of polygons (from: https://en.wikipedia.org/wiki/Polygon#/media/File:Polygon_types.svg)

In both 2D and 3D, the majority of objects drawn in graphics are made up of polygons; from the rectangle used to define an onscreen button, to the thousands of triangles used to construct a 3D model. It is therefore important to your knowledge of working in this area to have a strong understanding of what they are, how to construct them, and how to manipulate them.

Here are some interesting and useful facts about polygons:

- A convex polygon has all interior angles add up to less than or equal to 180 degrees, for example, each angle in an equilateral triangle is 60 degrees.

- A regular convex polygon has edges that all have the same length and fits neatly inside a circle.

- The sum of the interior angles of a simple n-gon is (n-2) x 180 degrees.

- The sum of the exterior angles of a simple n-gon is always 360 degrees.

For more useful facts, check out the Wikipedia entry at en.wikipedia.org/wiki/Polygon. We will be examining polygons and their geometry quite a bit throughout the book as we begin developing game and graphics scenarios, however, we will have a quick look at how they are dealt with in Pygame.

Let's do it...

The only difference between drawing a line and a polygon with Pygame is the addition of more than two points that define the corners of the shape. Try this out now:

1. Create a new Python file called `Polygons.py` and add in the usual starter code for drawing a window.

2. Before the `while` loop, define a variable for the color white like this:

    ```
    white = pygame.Color(255, 255, 255)
    ```

3. Inside the `while` loop, immediately above the `update()` method definition, start typing `pygame.draw`. The context sensitive help in PyCharm will reveal the types of things you can draw; in the list, you will see `polygon`.

4. Now complete the line with the following:

    ```
    pygame.draw.polygon(screen, white, ((150,200),
                    (600,400),(400,600)))
    ```

    Here, take special note of where all the brackets are positioned. Each x, and y coordinate is encased in brackets and then all three coordinates are also inside brackets. This is essential to pass the three coordinates to the `polygon()` method.

5. Run the program. You will see a triangle like the one in *Figure 2.10*:

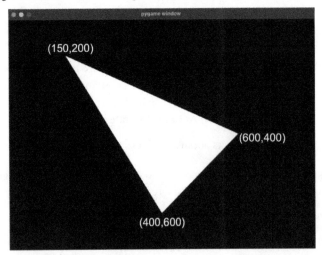

Figure 2.10: A drawing of a polygon with the vertex coordinates marked

The required parameters for the shape are the surface the polygon is being drawn on, the color of the polygon, and finally, a list of the vertices.

Your turn...

Exercise D: Instead of hardcoding the vertices of the polygon into the program, create a program that uses mouse clicks (like you did for drawing a line) to define the three points of a triangle.

Whether it is lines or polygons, all drawn items end up on the screen as pixels. In the case of these items, they have a mathematical structure that determines how the pixels are colored. The next element of a graphics image is already represented by pixels. Raster images are particularly useful for times that mathematics won't suffice. For example, using real photographs, placing textures onto 3D objects, and representing background images.

Raster Images

The last of the graphics elements we are going to examine in this chapter are raster images. These are the regular types of images used with computers such as those produced by displaying *PNG* or *JPG* files. They are presented as a grid of pixels where each pixel can be seen when the image is zoomed in, as illustrated in *Figure 2.11*:

Figure 2.11: A raster image of Smudge and a closeup of the corner of his eye showing individual pixels

If an image has a high resolution, you won't be able to make out each pixel.

As mentioned in our discussion of creating text, in Pygame, images can be drawn onto surfaces that don't appear on the main display until required. This also goes for raster images.

First, they are loaded into memory and then you can put them onto the display over and over again at different locations using `screen.blit`. The image you choose can include transparency. This is a nice feature as it allows you to draw a sprite over the top of an existing background.

Blitting is a common technique used in graphics along with Boolean operations to place one image on top of another by comparing pixel values. For a thorough explanation, I would encourage you to read en.wikipedia.org/wiki/Bit_blit. In the following exercise, we will employ the blit operation to place a background image in a window and overlay a sprite.

Let's do it...

In this exercise, we will work with a background image for the entire window, as well as one for a sprite. To do this, follow these steps:

1. Create a new Python file called ShowRaster.py and add the usual starting code.

2. Browse the web for a background image to fill the entire window. Whatever the size of the image you select, you can adjust the size of the window so it's a neat fit.

3. Drag and drop that image file into PyCharm in the same manner that you did for the font (i.e., hold down the *Ctrl* key to ensure it is copied across). If you like, you can also create a separate folder called images to place image files within.

4. Now modify your code to include the script to load and display the image like this:

```python
import pygame


pygame.init()
screen_width = 800
screen_height = 400
screen = pygame.display.set_mode((screen_width,
                                  screen_height))


#Add a title to the display window
pygame.display.set_caption('A Beautiful Sunset')
done = False

#load background image
background =
    pygame.image.load('images/background.png')
while not done:
    for event in pygame.event.get():
        if event.type == pygame.QUIT:
            done = True
    screen.blit(background, (0, 0))
```

```
    pygame.display.update()
pygame.quit()
```

To begin, I have made the window the same size as the image I chose, which is 800 x 400. Next, I've included the `set_caption` method to show you how to change the title of the window. The image itself is loaded into the display called `background` with the `pygame.image.load` method. Note, this method takes a string as the path to the image.

You can find this path using the same method we employed for the TrueType text file. The loading of the image happens outside the `while` loop as it only needs to occur once. The background image is then blitted onto the screen with its upper left-hand corner aligned at (0, 0). *Figure 2.12 (a)* illustrates an image used as the background of a window:

(a)   (b)

Figure 2.12: The background image (a) and with a sprite (b)

5. We are now going to blit another image with a transparent part over the top of the background. For this, you will require a PNG image like something that you might use as a 2D game character. You can find such an image at www.iconarchive.com. Download one you like the look of at a resolution of 64 x 64 and import it into PyCharm.

6. The process to add the *PNG* file to the window is the same as for adding the background. First, the image is placed into a variable outside of the `while` loop and then the image is blitted onto the surface. Placing the code for the new image after the background image will ensure the second image is on top, as you can see in *Figure 2.12 (b)*. The partial listing of this code showing the line additions in bold is as follows:

```
background =
    pygame.image.load('images/background.png')
sprite =
    pygame.image.load('images/Bird-blue-icon.png')
while not done:
    for event in pygame.event.get():
        if event.type == pygame.QUIT:
```

```
        done = True
    screen.blit(background, (0, 0))
    screen.blit(sprite, (100, 100))
    pygame.display.update()
pygame.quit()
```

The `load()` method puts the sprite into memory where the sprite name is an actual display. It just hasn't been drawn in the window. That's the job of the following `blit()` method.

Despite all the clever mathematics that computer scientists have devised throughout the years, often the image quality and complexity is just sometimes better and easier represented as a raster image. There are many algorithms out there and some that we will explore in this book that can be used to procedurally generate pretty convincing images. But sometimes, they aren't necessarily a good substitute for the real thing.

Summary

We've now covered the fundamental aspects of computer graphics. With all these elements, anything can be drawn on the screen. Although the examples we have worked with have been exclusively 2D, the very same principles are used throughout 3D, as you will discover while working through this book. You now have enough knowledge to construct your own simple 2D graphics projects that integrate external images and fonts and make use of mouse interactions.

In *Chapter 3, Line Plotting Pixel by Pixel*, we will take a look behind the scenes of the higher-level drawing methods that have been examined in this chapter, with a step-by-step look at an algorithm that has been used since the birth of computer graphics to process how individual pixels should be colored when drawing a line.

Answers

Exercise A:

```
screen.set_at((x, y), pygame.Color(0, int(x/1000 * 255),
                    int(y/800 * 255)))
```

Exercise B:

```
y = 8x + 0.5
```

Exercise C:

For another line of text with a different font, you will need to declare two fonts and two texts like this:

```
font1 = pygame.font.Font('fonts/Authentic Sheldon.ttf', 120)
font2 = pygame.font.Font('fonts/PinkChicken-Regular.ttf', 120)
text1 = font1.render('Penny de Byl', False, white)
text2 = font2.render('Another line of text', False, white)
```

Then, inside the `while` loop, blit both:

```
screen.blit(text1, (10, 10))
screen.blit(text2, (10, 100))
```

Exercise D:

```
import pygame
from pygame.locals import *

pygame.init()
screen_width = 800
screen_height = 800
screen = pygame.display.set_mode((screen_width, screen_height))

pygame.display.set_caption('Triangle Clicker')
done = False
white = pygame.Color(255, 255, 255)
timesClicked = 0
pygame.font.init()

while not done:
  for event in pygame.event.get():
    if event.type == pygame.QUIT:
      done = True
    elif event.type == MOUSEBUTTONDOWN:
      if timesClicked == 0:
        point1 = pygame.mouse.get_pos()
      elif timesClicked == 1:
        point2 = pygame.mouse.get_pos()
```

```
        else:
          point3 = pygame.mouse.get_pos()
        timesClicked += 1
        if timesClicked > 2:
          pygame.draw.polygon(screen, white,
                    (point1, point2, point3))
          timesClicked = 0

  pygame.display.update()
pygame.quit()
```

3
Line Plotting Pixel by Pixel

It's difficult to appreciate what goes on mathematically behind the scenes of high-level drawing methods, such as those used by Pygame methods, unless you've written a program to plot objects pixel by pixel – because when it comes down to it, this is how the methods produce their rendered results. To that end, we are going to explore a little deeper the techniques involved in drawing a line pixel by pixel onto the screen. At first, you might assume this is quite a simple thing to do, but you'll soon discover that it's quite complex.

As we progress through the content, it will become clear why special algorithms are required to draw mathematical constructs such as lines on a pixel display. The algorithm we are going to explore is also a key technique that can be used for various other graphics and game-related functions wherever straight-line calculations are needed.

Once you've completed this chapter, you will have a new algorithm in your toolkit that you can employ whenever a line needs to be drawn and you can't rely on out-of-the-box API calls.

In this chapter, we will explore line plotting by looking at the following:

- The Naïve Way – Drawing a line with brute force
- The improved Approach: Using Bresenham's algorithm
- Drawing Circles the Bresenham Way
- Anti-aliasing

Technical requirements

In this chapter, we will be using Python, PyCharm, and Pygame, as used in previous chapters. Before you begin coding, create a new folder in the PyCharm project for the contents of this chapter.

The solution files containing the code can be found on GitHub at `https://github.com/PacktPublishing/Mathematics-for-Game-Programming-and-Computer-Graphics/tree/main/Chapter03`.

The Naïve Way: Drawing a line with brute force

So, you have an equation to draw a line and want to plot it on the screen pixel by pixel. The naïve approach would be to loop through a series of x values, calculate what the y value should be at that x location given the equation, and then plot x, y as a colored pixel. When you do, a very unfortunate issue will arise.

An issue occurs because pixel locations are represented as whole numbers; however, points on a straight line can in fact be floating points. Consider the line in *Figure 3.1*, showing a grid of pixel locations and a line drawn through them:

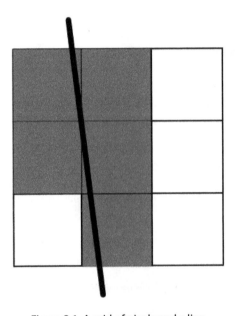

Figure 3.1: A grid of pixels and a line

If you were to draw every pixel the line touched (shown by the shaded squares), it wouldn't give a very accurate presentation of the line. In addition, the real location of any point on the line could calculate to a floating-point value. When this occurs, the value is rounded to an integer to find the closest possible pixel. However, this causes other issues that produce odd gaps in the drawing.

You are about to experience this issue first-hand in an exercise designed to first show you the wrong way to draw a line. Then, we'll look at a better way.

So, let's begin by taking the naïve approach to drawing a line pixel by pixel.

Let's do it...

The best way to understand how to manually draw a line pixel by pixel is to program it yourself. So, let's get started:

1. Create a new Python script called `LinePlot.py` in PyCharm and try the following code:

```python
import pygame
pygame.init()
screen_width = 1000
screen_height = 800
screen = pygame.display.set_mode((screen_width,
                                  screen_height))
done = False
white = pygame.Color(255, 255, 255)
green = pygame.Color(0, 255, 0)
xoriginoffset = int(screen.get_width() / 2)
yoriginoffset = int(screen.get_height() / 2)
while not done:
  for event in pygame.event.get():
    if event.type == pygame.QUIT:
      done = True
  # x axis
  for x in range(-500, 500):
    screen.set_at((x + xoriginoffset, yoriginoffset),
                  green)
  # y axis
  for y in range(-400, 400):
    screen.set_at((xoriginoffset, y + yoriginoffset),
                  green)
  for x in range(-500, 500):
    y = 2 * x + 4 #LINE EQUATION
    screen.set_at((x + xoriginoffset, y +
                   yoriginoffset), white)
  pygame.display.update()
pygame.quit()
```

This is plotting the same original line equation we used in the DESMOS exercise in *Chapter 2, Let's Start Drawing* (you can see it in *Figure 2.3*, albeit upside down because y is flipped). The `xoriginoffset` and `yoriginoffset` values are used to place the origin in the center of the window. You will notice that there are also green pixels drawn to show the x and y axes. The result is shown in *Figure 3.2*:

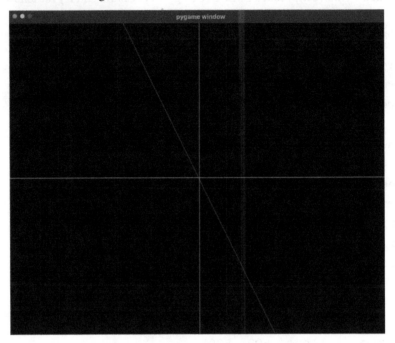

Figure 3.2: A plot of y = 2x + 4 in Pygame

Besides the image being upside down in the Python render, the line is the same as the one drawn in DESMOS. Or is it? With high-resolution monitors, it can be difficult to see the issue that arises from plotting a line in this way. So, let's try something else.

2. Change the equation of the line (indicated by the `LINE EQUATION` comment in the previous code) to the following:

```
y = int(0.05 * x) - 100
```

The value of `100` at the end is the y intercept and as such (because the y axis is flipped), the line will cross the y axis further up the screen.

But what of `int(0.05 * x)`? Well, `0.05` is the slope, which in this case is very flat. The typecast to an integer by wrapping the value with brackets and typing `int` at the front are to appease the `set_at()` method. Why? Well, let's consider something. The window is 1,000 pixels wide by 800 pixels high. If you wanted to plot a point at pixel (10, 10), I'm sure you can determine where that would appear. However, what if you wanted to plot a point at (10.5, 10)?

Where is the pixel with the x coordinate of 10.5? There is none. Pixels are counted in integers (whole numbers) only. So, while you can generate numbers that aren't integers, you can't plot them on a computer display as a pixel.

As such, the resulting line that you will see, zoomed in, looks like that in *Figure 3.3 (a)*. Notice the stepping in the line that takes place? This is the result of rounding to integers for pixel locations:

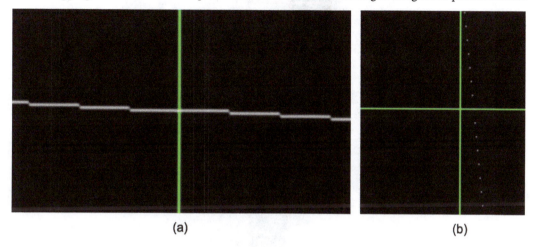

(a)   (b)

Figure 3.3: A close-up of an integer line render – (a) y = 0.5x – 100 and (b) y = 10x – 100

3. Now try `y = (10 * x) - 100`. It reveals the gaps that are left after integer pixel plotting with a steep slope in the y direction, as shown in *Figure 3.3 (b)*.

This might not seem like a big deal until you want to draw a straight line between two points in a game environment and want them to be connected. A high screen resolution is fine, but if you've got a grid map for a real-time strategy game and want to connect the grid locations with a line for, say, building a road, then you need **Bresenham's algorithm**.

Enter Bresenham: The improved approach

Jack Bresenham developed this algorithm while working for IBM in 1962. In short, it plots a straight line but doesn't leave pixels disconnected from each other. The basic premise is that when drawing a line pixel by pixel, each successive pixel must be at least a distance of 1 in either the x or y (or both) direction. If not, a gap will appear. While the naïve approach we've just used creates a plot where all the x values are a distance of 1 apart, the same isn't always true for the y values. The only way to ensure all x and y values are a distance of 1 apart is to incrementally plot the line from point 1 to point 2, ensuring that either the x or y values are changing by a maximum of 1 with each loop.

Consider the close-up of a steep line being plotted in *Figure 3.4*:

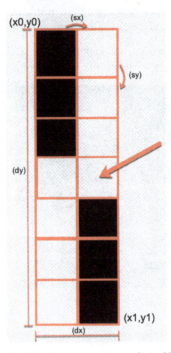

Figure 3.4: A line being constructed pixel by pixel

The values for dx (change in x values) and dy (change in y values) represent the horizontal pixel count that the line inhabits and dy is that of the vertical direction. Hence, dx = abs(x1 – x0) and dy = abs(y1 – y0), where abs is the absolute method and always returns a positive value (because we are only interested in the length of each component for now). In *Figure 3.4*, the gap in the line (indicated by a red arrow) is where the x value has incremented by 1 but the y value has incremented by 2, resulting in the pixel below the gap. It's this jump in two or more pixels that we want to stop.

Therefore, for each loop, the value of x is incremented by a step of 1 from x0 to x1 and the same is done for the corresponding y values. These steps are denoted as sx and sy. Also, to allow lines to be drawn in all directions, if x0 is smaller than x1, then sx = 1; otherwise, sx = -1 (the same goes for y being plotted up or down the screen).

With this information, we can construct pseudo code to reflect this process, as follows:

```
plot_line(x0, y0, x1, y1)
    dx = abs(x1-x0)
    sx = x0 < x1 ? 1 : -1
    dy = -abs(y1-y0)
    sy = y0 < y1 ? 1 : -1
```

```
while (true)   /* loop */
  draw_pixel(x0, y0);
  #keep looping until the point being plotted is at x1,y1
  if (x0 == x1 && y0 == y1) break;
  if (we should increment x)
    x0 += sx;
  if (we should increment y)
    y0 += sy;
```

The first point that is plotted is x0, y0. This value is then incremented in an endless loop until the last pixel in the line is plotted at x1, y1. The question to ask now is: "How do we know whether x and/or y should be incremented?"

If we increment both the x and y values by 1, then we get a 45-degree line, which is nothing like the line we want and will miss its mark in hitting (x1, y1). The incrementing of x and y must therefore adhere to the slope of the line that we previously coded to be m = (y1 - y0)/(x1 - x0). For a 45-degree line, m = 1. For a horizontal line, m = 0, and for a vertical line, m = ∞.

If point1 = (0,2) and point2 = (4,10), then the slope will be (10-2)/(4-0) = 2. What this means is that for every 1 step in the x direction, y must step by 2. This of course is what is creating the gap, or what we might call the error, in our line-drawing algorithm. In theory, the largest this error could be is dx + dy, so we start by setting the error to dx + dy. Because the error could occur on either side of the line, we also multiply this by 2. As each pixel is calculated, the error is updated with the values of dy if incrementing x and dx if incrementing y. Whether or not x and y are incremented then becomes a matter of testing the error against the actual dx and dy values that specify the x and y components of the line's slope.

For the full derivation of these calculations, those who are interested are encouraged to read *Section 1.6* of members.chello.at/~easyfilter/Bresenham.pdf. In the following exercise, you will get a sense of the issues involved with using a linear equation to plot lines and experience first-hand pixel skipping before modifying the code to include Bresenham's algorithm.

Let's do it...

We will now examine Bresenham's algorithm more closely by using it for a very common problem that comes up in graphics programming: drawing a line on the screen between two points. The two points in this case will come from the mouse. We will begin by illustrating the line-drawing issue using a typical linear equation, before adding Bresenham's improvements. Try this out:

1. Create a new Python script file called PlotLineNaive.py and start it with the following:

    ```
    import pygame
    from pygame.locals import *
    ```

```python
pygame.init()
screen_width = 1000
screen_height = 800
screen = pygame.display.set_mode((screen_width, screen_
height))
done = False
white = pygame.Color(255, 255, 255)
green = pygame.Color(0, 255, 0)
timesClicked = 0
while not done:
    for event in pygame.event.get():
        if event.type == pygame.QUIT:
            done = True
        elif event.type == MOUSEBUTTONDOWN:
            if timesClicked == 0:
                point1 = pygame.mouse.get_pos()
            else:
                point2 = pygame.mouse.get_pos()
            timesClicked += 1
            if timesClicked > 1:
                pygame.draw.line(screen, white,
                                 point1, point2, 1)
                timesClicked = 0
    pygame.display.update()
pygame.quit()
```

2. Add the following new method before the `while` loop:

```python
times_clicked = 0
def plot_line(point1, point2):
  x0, y0 = point1
  x1, y1 = point2
  m = (y1 - y0)/(x1 - x0)
  c = y0 - (m * x0)
  for x in range(x0, x1):
    y = m * x + c # LINE EQUATION
    screen.set_at((int(x), int(y)), white)
```

```
while not done:
```

This uses the two points to calculate the slope (m) and the y intercept (c) derived from the equation for a line, and then loops across in the x direction between the x values of each point. The value of y is calculated using the line equation.

3. To use this method to plot a line, locate the following lines:

```
pygame.draw.line(screen, white, point1, point2, 1)
```

Replace it with the following:

```
plot_line(point1, point2)
```

4. Run the application and click twice to see the lines being drawn, as shown in *Figure 3.5*:

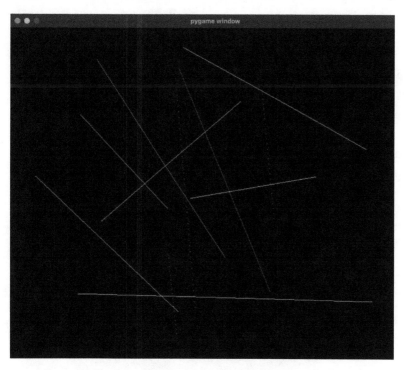

Figure 3.5: The lines drawn using a naïve approach to line plotting

You will immediately notice, besides the gaps in the lines with a steep slope, that you can only plot a line when point 1 is to the left of point 2. This is due to the for loop, which requires x0 to be smaller than x1. If you want to be able to draw in either direction, you can add a small test to swap the points around at the very top of the plot_line method, like this:

```
if point2[0] < point1[0]:
    temp = point2
```

```
        point2 = point1
        point1 = temp
```

Now it's time to implement Bresenham's algorithm in our project with Python.

5.   Modify the code for the `plot_line` method to this:

```
def plot_line(point1, point2):
  x0, y0 = point1
  x1, y1 = point2
  dx = abs(x1 - x0)
  if x0 < x1:
    sx = 1
  else:
    sx = -1
  dy = -abs(y1 - y0)
  if y0 < y1:
    sy = 1
  else:
    sy = -1
  err = dx + dy
  while True:
    screen.set_at((x0, y0), white)
    if x0 == x1 and y0 == y1:
      break
    e2 = 2 * err
    if e2 >= dy:
      err += dy
      x0 += sx
    if e2 <= dx:
      err += dx
      y0 += sy
```

Once you have entered this into PyCharm and tried it out, you will find that nicely connected lines can be drawn for all slopes between the first and second clicks, as shown in *Figure 3.6*:

Figure 3.6: The lines drawn with Bresenham's algorithm

You might think we have over-complicated or over-explained this code for a book on computer graphics and games; however, I cannot emphasize enough how fundamental this algorithm is to your understanding of how objects are drawn on the screen and how often when you are creating game mechanics you will want a player to draw a straight line across an integer-based grid and produce a connected straight line. Drawing a straight line in a 2D game engine nowadays might be an easy procedure thanks to APIs, but in 3D, it becomes something completely different. Bresenham's line algorithm has been a go-to for me over the past 20 years, as it not only is applicable to drawing lines pixel by pixel but also will work in any grid-type system where you need to plot a straight line from one cell to another without skipping any cells in between. This is particularly useful in artificial intelligence dynamics in games where you require characters to move across a grid, or even in city-building simulations that can also be played in a grid.

Bresenham didn't end his exploration of drawing lines with pixel grids; he also investigated other shapes. Let's take a look at the circle-drawing algorithm.

Drawing circles the Bresenham way

Drawing circles pixel by pixel can also suffer from the same issue of pixel skipping if the naïve approach of using a circle equation to calculate pixel locations is used. Consider the implicit circle equation:

$$y = \pm\sqrt{radius^2 - x^2}$$

The naïve approach to drawing this with pixels could be achieved with the following code:

```python
import math
import pygame
pygame.init()
screen_width = 400
screen_height = 400
screen = pygame.display.set_mode((screen_width,
                                  screen_height))
pygame.display.set_caption('Naive Circle')
done = False
white = pygame.Color(255, 255, 255)
radius = 50
center = (200, 200)
while not done:
    for event in pygame.event.get():
        if event.type == pygame.QUIT:
            done = True
    for x in range(-50, 50):
        y = math.sqrt(math.pow(radius, 2) - math.pow(x, 2))
        screen.set_at((int(x + center[0]),
                    int(y + center[1])),
                    white)
        y = -math.sqrt(math.pow(radius, 2) - math.pow(x,
                                                    2))
        screen.set_at((int(x + center[0]),
                    int(y + center[1])), white)
    pygame.display.update()
pygame.quit()
```

You may type this into PyCharm to see it in action if you like. When you do, the result will show missing pixels when the slope of the circle gets closer to vertical, as shown in *Figure 3.7*:

Figure 3.7: A naively plotted circle

The naïve circle-drawing approach suffers from exactly the same issue as drawing steep-sloped lines; pixels go missing when floats are rounded to integers. To alleviate this, Bresenham came up with a midpoint circle algorithm that determines the best pixel to select to draw based on how a floating value sits between two pixels. Consider the diagram in *Figure 3.8*; the cross lines are pixel locations:

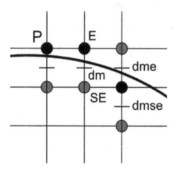

Figure 3.8: The midpoint circle algorithm points of choice

The point **P** is a pixel chosen to be used as a point on the circle as the actual equation for the circle comes close to this coordinate. If a point on a curve sits between pixels, the algorithm considers where the point is with respect to the midpoint (**dm**) between the pixel above the curve (**E**) and the point below the curve (**SE**). These are named as such because with respect to **P**, **E** is east and **SE** is southeast.

The algorithm for the midpoint circle is not dissimilar to Bresenham's line algorithm, as you will see in the following exercise.

Let's do it...

This exercise begins with the essential code to produce a circle with Bresenham's midpoint circle algorithm. Follow along:

1.   Create a new Python file called `BresenhamCircle.py` and add the following code:

```python
import pygame
from pygame.locals import *  #support for getting
                             #mouse button events
pygame.init()
screen_width = 400
screen_height = 400
screen = pygame.display.set_mode((screen_width,
                                   screen_height))
done = False
white = pygame.Color(255, 255, 255)
def plot_circle(radius, center):
    x = 0
    y = radius
    d = 5/4.0 - radius
    circle_points(x, y, center)
    while y > x:
        if d < 0:  # select E
            d = d + 2 * x + 3
            x += 1
        else:  # select SE
            d = d + 2 * (x - y) + 5
            x = x + 1
            y = y - 1
        circle_points(x, y, center)
while not done:
    for event in pygame.event.get():
        if event.type == pygame.QUIT:
            done = True
    plot_circle(50, (200, 200))
    pygame.display.update()
pygame.quit()
```

This won't run yet as we haven't created the `circle_points()` method.

The value of d here is a decision variable taking on the value of the circle at the midpoint. In *Figure 3.8*, d is denoted by **dm**. Its value relative to **P** will be the following:

$$dm = (P_x + 1, P_y - 1/2) = (P_x + 1)^2 + (P_y - 1/2)^2 - R^2$$

If the value of d is less than 0, then the line is above the midpoint and the pixel **E** is chosen. That makes the calculation for the next midpoint, dme, as follows:

$$dme = (P_x + 2, P_y - 1/2) = (P_x + 2)^2 + (P_y - 1/2)^2 - R^2 = dm + (2P_x + 3)$$

However, if it were a **SE** pixel, then the next value of dm would be at dmes, which would equate to the following:

$$dmes = (P_x + 2, P_y - 3/2) = (P_x + 2)^2 + (P_y - 3/2)^2 - R^2 = dm + (2P_x - 2P_y + 5)$$

Both the dme and dmes equations can be seen in the code as being used to update the decision value for the next x value.

If you were to run this code as is and just draw one pixel for each (x, y) combo calculated, you'd find that it only draws points in the first octant of the circle (highlighted in *Figure 3.9*). But lucky for us, a circle has eight points of symmetry around its center. That means we can simply plot all the combinations of x and y around the circle's origin to cover all the points.

Let's get back to the code.

2. Before the plot_circle() method, add a new method to draw all the points on the circle, as follows:

```
white = pygame.Color(255, 255, 255)
def circle_points(x, y, center):
    screen.set_at((x + center[0], y + center[1]),
                    white)
    screen.set_at((y + center[0], x + center[1]),
                    white)
    screen.set_at((y + center[0], -x + center[1]),
                    white)
    screen.set_at((x + center[0], -y + center[1]),
                    white)
    screen.set_at((-x + center[0], -y + center[1]),
                    white)
    screen.set_at((-y + center[0], -x + center[1]),
                    white)
    screen.set_at((-y + center[0], x + center[1]),
                    white)
```

```
       screen.set_at((-x + center[0], y + center[1]),
                        white)
    def plot_circle(radius, center):
```

This new method uses the Pygame pixel-plotting method to draw eight points around the perimeter with each (x, y) value calculated. I've also added an offset for the center as the algorithm creates points centered around (0, 0); however, if you want to move the circle around on the screen, it requires a new origin. In this case, it is being moved by (200, 200) into the center of my window.

Here is the result of the code:

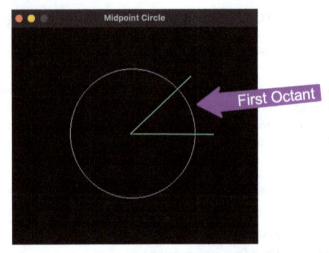

Figure 3.9: The result of the midpoint circle algorithm with the first octant highlighted

While the Pygame API makes drawing lines and circles straightforward, now having examined Bresenham's algorithms at a pixel-by-pixel level, you'll have a deeper understanding of the mathematics that is going on behind the scenes.

However, the lines and circles we have produced in this section look rather jagged close up. The reason for this will now be revealed and solutions posed to create smoother lines.

Anti-aliasing

In the previous sections, we worked on producing pixel-by-pixel lines where one pixel was chosen over another in order to create a continual line or circle. If you zoom in on these pixels, however, you will find that they are quite jagged in appearance, as illustrated by the zoomed-in section of a circle shown in *Figure 3.10*:

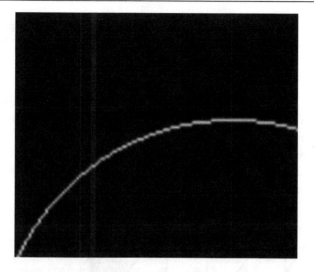

Figure 3.10: A zoomed-in section of the Bresenham circle pixels

This effect occurs because of the way that integer values are chosen to represent the drawing. However, when we looked at the actual equations for lines and circles in the *The Naïve Way: Drawing a line with brute force* and *Drawing Circles the Bresenham way* sections, it was clear they involved floating-point values and seemed to pass through more than one pixel. To improve the look of these kinds of drawings, a process called anti-aliasing is employed to blur the pixels that are neighbors of the line pixels to fade the edges of the line. *Figure 3.11* shows the zoomed-in edge of a black shape that is aliased (*Figure 3.11(a)*) and anti-aliased (*Figure 3.11(b)*):

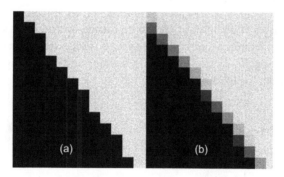

Figure 3.11: A zoomed-in section of a shape without (a) and with (b) anti-aliasing

There are several techniques for achieving anti-aliasing that make rasterized digital images appear a lot smoother, including multisampling, supersampling, and fast approximation. For further details on these algorithms, refer to en.wikipedia.org/wiki/Anti-aliasing.

In short, however, these algorithms work by considering the color of their neighboring pixels and averaging out the colors between them to create a blur or *tweening* effect between the two. In the case of drawing black pixels on a white background, wherever the contrast difference between the two is greatest, gray pixels are used to smooth out the jagged edge.

In the case of geometry, instead of considering whether a pixel is part of the shape or not, that is, considering it true *if it is a part of the line* and false *if it isn't part of the line*, it's given a weight based on how much of the line covers it. This is shown in *Figure 3.12* with a close-up of a line segment that is 1 pixel wide, as defined by the rectangle in *Figure 3.12(a)*. In *Figure 3.12(b)*, the pixels mostly covered by the line are black and then, depending on how much of the line covers the other pixels, they are weighted with lesser values of gray:

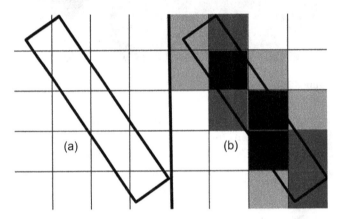

Figure 3.12: Anti-aliasing according to weighting pixels

In most engines, anti-aliasing is an option that can be turned on and off. It's usually applied as a postproduction effect, and therefore requires the image to be processed twice: once for drawing and a second time for anti-aliasing. If you are interested in drawing and comparing an anti-aliased circle with Pygame, you can try out the following code:

```
from pygame import gfxdraw
# Without anti-aliasing
pygame.draw.circle(screen, white, (500, 500), 200, 1)
# With anti-aliasing
pygame.gfxdraw.aacircle(screen, 500, 500, 210, white)
```

Anti-aliasing will make a big difference to the visual quality of your graphics applications, so be sure to look out for it as a setting in any software you are using.

Summary

In this chapter, we've created code to force the errors incurred when trying to plot floating points in whole-number pixel locations. Mathematically speaking, the focus of the chapter was on Bresenham's algorithms for drawing lines and circles that take the respective equations and create a raster image from them while working with integer values to ensure a pixel-perfect drawing.

In *Chapter 4, Graphics and Game Engine Components*, we will explore the structure of graphics engines and look at how they are put together to cater to the numerous functions they need to serve.

4

Graphics and Game Engine Components

No matter which graphics or games development engine you examine, they are built using a very similar architecture, the reason being that they all have to work with the same graphics devices, **central processing units (CPUs)**, and **graphics processing units (GPUs)** to produce stunning visuals with exceptional frame rates. If you consider the nature of graphical elements that are drawn on a screen, they are also constructed from numerous layers depending on their functionality, dimensions in pixels, coloring, animation, lighting, physics, and user interaction.

In this chapter, we will begin exploring some typical application architectures as we begin to build our own. To enable the drawing of **three-dimensional (3D)** objects in Python we will use the **PyOpenGL** package that implements **OpenGL** (a 3D graphics library). You will learn how each of the graphics components works together to create a scalable and flexible graphics engine.

Herein, we will cover the following topics:

- Exploring the OpenGL Graphics Pipeline
- Drawing Models with Meshes
- Viewing the Scene with Cameras
- Projecting Pixels onto the Screen
- Understanding 3D Coordinate Systems in OpenGL

Technical requirements

In this chapter, we will be using Python, PyCharm, and Pygame, as used in previous chapters.

Before you begin coding, create a new folder in the PyCharm project for the contents of this chapter, called Chapter_4.

You will also need PyOpenGL, the installation of which will be covered during the first exercise.

The solution files containing the code can be found on GitHub at `https://github.com/PacktPublishing/Mathematics-for-Game-Programming-and-Computer-Graphics/tree/main/Chapter04`.

Exploring the OpenGL Graphics Pipeline

In software, an engine is an application that takes all the hard work out of creating an application by providing an **application programming interface (API)** specific to the task. **Unity3D**, for example, is a game engine. It's a tool for creating games and empowers the programmer by removing the need to write low-level code.

An engine consists of several modules dedicated to tasks, as shown in *Figure 4.1*:

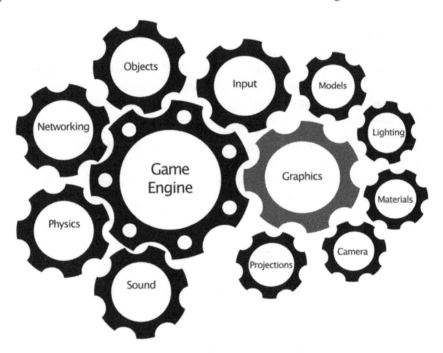

Figure 4.1: Typical components of a game engine

A mainstream graphics application needs to possess all the fundamental abilities to display and move graphics elements (such as the ones we examined in *Chapter 2, Let's Start Drawing*) in either 2D or 3D, with most 3D engines able to display 2D simply by removing the z axis. These elements work together to take a 2D or 3D model through the graphics pipeline where it is transformed from a set of vertices into an image that appears on the screen. The graphics pipeline includes several highly mathematical steps to achieve the final product. These steps are illustrated in the following diagram:

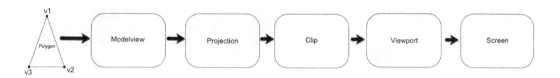

Figure 4.2: Graphics pipeline

Let's look at each of these steps here:

- First, the vertices of the model to be drawn are transformed by the `modelview` into camera space (also called **eye space**). This changes the model's vertices, edges, and other characteristics relative to the camera's location and other properties. At this point, the model color information is attached to each vertex.

- The vertices and colors are then passed to the projection, which transforms the values into a view volume. At this point in the pipeline, the model has been reduced to pixels.

- **Clipping** then discards any drawing information that is outside the viewing volume of the camera as it is no longer required. It can't be seen, so why draw it?

- Then, the viewport takes the pixel information and transforms it yet again into device-specific coordinates of the output device, and the pixel information is seen on the screen.

We will revisit sections of the graphics pipeline as we move forward with examples that will assist your understanding.

First, however, we are going to set up PyCharm with OpenGL to allow us to create more advanced 3D graphics programs as we move forward. OpenGL is a cross-platform, freely available API for interacting with the GPU to achieve hardware-accelerated drawing and has been an industry standard since the early 1990s.

Let's do it...

In this exercise, we will walk through the process to get Python's OpenGL package installed and tested. Before you begin, as already mentioned, create a new folder in PyCharm to store the code from this chapter. Then, complete these steps:

1. In PyCharm, go to the **Python Packages** window and install PyOpenGL (in the same way we installed Pygame). The correct package to install is shown in *Figure 4.3*:

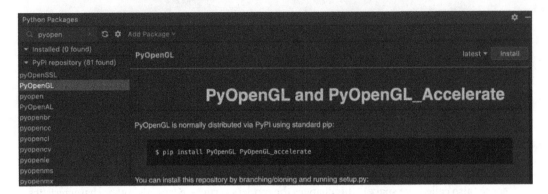

Figure 4.3: Installing PyOpenGL in PyCharm

2. Create a new Python file called `OpenGLStarter.py` and add the following code to it:

```python
import pygame
from pygame.locals import *
from OpenGL.GL import *
pygame.init()
screen_width = 500
screen_height = 500
screen = pygame.display.set_mode((screen_width,
                                  screen_height),
                                 DOUBLEBUF|OPENGL)
pygame.display.set_caption('OpenGL in Python')
done = False
white = pygame.Color(255, 255, 255)
while not done:
  for event in pygame.event.get():
    if event.type == pygame.QUIT:
      done = True
  glClear(GL_COLOR_BUFFER_BIT | GL_DEPTH_BUFFER_BIT)
  pygame.display.flip()
pygame.quit()
```

When run, this code will open up a window as before. Note that much of what we've already been working with is still the same; however, the noticeable differences are set out here:

- DOUBLEBUF|OPENGL option in set_mode(): We have moved from drawing a single image buffer to a double buffer and asked Pygame to use OpenGL to perform the rendering. A double buffer is required when drawing animated objects, as we will be doing later in this book. It means that one image is on the screen while another is being drawn behind the scenes.

- glClear() inside the while loop: glClear() clears the screen using the GL_COLOR_ BUFFER_BIT | GL_DEPTH_BUFFER_BIT bitmask, which removes color and depth information.

- pygame.display.flip(): Here, flip() replaces the update() method we previously used and switches the buffer images so that the background buffer is sent to the screen.

For in-depth information on any OpenGL methods introduced, the interested reader is encouraged to visit the following web page:

https://www.khronos.org/registry/OpenGL-Refpages/gl4/

This script will now become the new starter file for our projects. Whenever a new Python project is started, be sure to copy this code into the new file unless otherwise directed.

With the basic program that we will be working with developed, it's time to look at some graphics/ game engine modules, in turn adding their basic functionality to our Python application as we proceed.

Drawing Models with Meshes

A model is an object drawn by the graphics engine. It contains a list of vertices that define its structure in terms of polygons. The collection of these connected polygons is known as a **mesh**. Each polygon inhabits a plane—this means that it is flat. A basic model with elementary shading appears faceted, such as that shown in *Figure 4.4*. This flat nature is hidden using differing materials, as we will discuss shortly:

Figure 4.4: A basic polygon mesh showing the flatness of each polygon

A **polygon mesh** is stored internally as a list of vertices and triangles. Triangles are chosen to represent each polygon over that of a square, as triangles require less storage and are faster to manipulate as they have one less vertex.

A typical data structure to hold a mesh is illustrated in the following diagram:

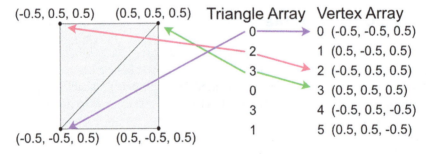

Figure 4.5: A vertex and triangle array used to define the triangles of a square

The example given stores a square split into two triangles. The vertex array holds all the vertices of the shape. Another array stores a series of integers that are grouped in threes and store the three vertices of a triangle. You can see how the first triangle (group of three values) in the triangle array references the indices of the vertex array, thus allowing a polygon to be defined.

The code to store the data for a mesh as shown in *Figure 4.5* is straightforward, as you are about to discover.

Let's do it...

We will now define a mesh class and integrate it into our project, as follows:

1. Create a new Python file called `HelloMesh.py` and add in the basic OpenGL starter code we created in the previous exercise.

2. Create another Python file called `Mesh3D.py` and add the following code to it:

```python
from OpenGL.GL import *
class Mesh3D:
  def __init__(self):
    self.vertices = [(0.5, -0.5, 0.5),
             (-0.5, -0.5, 0.5),
             (0.5, 0.5, 0.5),
             (-0.5, 0.5, 0.5),
             (0.5, 0.5, -0.5),
             (-0.5, 0.5, -0.5)
             ]
    self.triangles = [0, 2, 3, 0, 3, 1]
  def draw(self):
    for t in range(0, len(self.triangles), 3):
        glBegin(GL_LINE_LOOP)
        glVertex3fv(
            self.vertices[self.triangles[t]])
        glVertex3fv(self.vertices[self.triangles[t +
                1]])
        glVertex3fv(self.vertices[self.triangles[t +
                2]])
        glEnd()
```

In the class constructor, the vertices and triangles are hardcoded, and we will come back and change these later to something more flexible. The `draw` method allows this class to draw out the given triangles as a series of triangles. It loops through three vertices at a time and uses `GL_LINE_LOOP` to connect each.

OpenGL draws in blocks, starting with `glBegin()` and ending with `glEnd()`. The `glBegin()` parameter tells OpenGL how to draw. In this case, it is a line that goes from vertex to vertex and then reconnects itself with the first vertex.

3. To see the drawing in action, modify `HelloMesh.py` thus:

```python
import pygame
from Mesh3D import *
from pygame.locals import *

..

mesh = Mesh3D()
while not done:
  for ..
  glClear(GL_COLOR_BUFFER_BIT | GL_DEPTH_BUFFER_BIT)
  mesh.draw()
  pygame.display.flip()

..
```

With these changes, you are adding a mesh into your project and then ensuring it is drawn in the main loop.

4. Run `HelloMesh.py` to see the result, as illustrated in *Figure 4.6*:

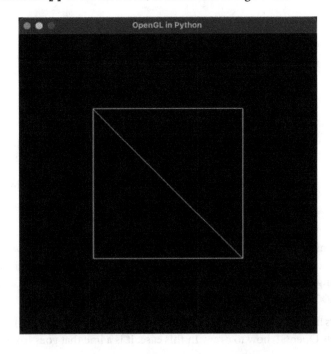

Figure 4.6: Drawing a square mesh with two triangles

Your turn...

Exercise A: Consider the mesh drawn in the previous section as the top of a cube. This cube extends from (-0.5, -0.5, -0.5) to (0.5, 0.5, 0.5). Define values for the vertices and triangles arrays, and create a class called `Cube` that inherits from `Mesh3D` and that will draw this new cubic mesh. (*Hint*: you only need to define the vertices and triangles; the `draw()` method will work as is.)

The code should begin like this:

```
from Mesh3D import *
class Cube(Mesh3D):
  def __init__(self):
    self.vertices = (insert vertices)..
    self.triangles = (insert triangles)..
```

Thus far in our project, we've not had a lot of influence over the location of the mesh within the 3D space or how we are looking at it. To allow for a better view of the cube or manipulate where it is appearing on the screen, we now need to consider the concept of cameras before adding them to the code.

Viewing the Scene with Cameras

The camera is responsible for taking the coordinates of objects (specifically, all their vertices and other properties) and transforming them in the `modelview` of *Figure 4.2*. If you consider the vertices of the square mesh we looked at in the previous section, its coordinates are not going to retain those values as it will be moved around in the graphics/game world. The `modelview` takes the world location of the mesh and the location of the camera, as well as some additional properties, and determines the new coordinates of the mesh so that it can be rendered correctly. Note that I am being purposefully abstract in describing this process right now as there's a lot of mathematics involved, and I don't want to complicate the discussion at this point!

The camera is a virtual object through which the world is visualized. It is the eye of the viewer and can be moved around the environment to provide different points of view. As illustrated in the following screenshot, the camera consists of a near plane that represents the screen where pixels are drawn, and a far plane that determines how far back into the world the camera can see:

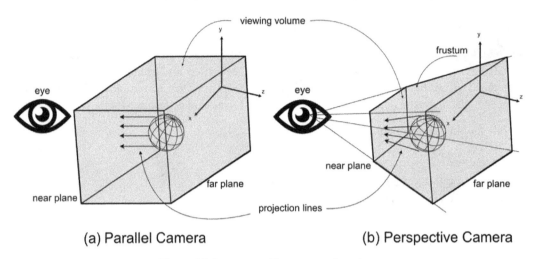

(a) Parallel Camera   (b) Perspective Camera

Figure 4.7: A camera with a perspective view

Any objects in front of the near plane or behind the far plane are culled. The area contained between the near and far planes is called the **viewing volume**.

The shape of the viewing volume differs between camera types and influences how a scene is rendered on the screen, as illustrated in *Figure 4.8*. Essentially, there are two camera types: **parallel** (also called **orthogonal**) and **perspective**. The shape of the viewing volume for a parallel camera is a rectangular prism, whereas, for perspective, it is called a **frustum**, which is a four-sided pyramid with the top cut off. This frustum can be seen in *Figure 4.8a*. In parallel, all objects inside the viewing volume are projected by straight lines at the near plane to be drawn. Unlike our own natural viewing of the real world where objects that are further away look smaller, in parallel, all objects retain their original scaling, no matter how far away they are from the eye. The other type of camera, shown in *Figure 4.8b*, gives a perspective view. This is closer to our own viewing outlook. This is illustrated in *Figure 4.8c*, where the scene containing several spheres that are all the same size but at different distances from the camera *(a)* appears to be all the same size in the parallel view of *(b)* but appears to look more natural in perspective *(c)*:

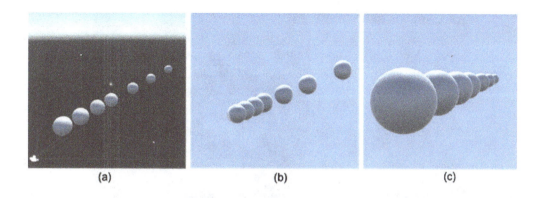

(a)   (b)   (c)

Figure 4.8: A 3D scene (a) viewed in parallel (b) and perspective (c)

When we look out into the world, for a person with normal eyesight, because our eyes are situated horizontally, we see the world as being wider than it is tall. If you were to hold your arms horizontally out from your sides while looking straight ahead and then slowly bring your arms around to your front, the moment that you can see your hands measures your **field of view** (**FOV**). For the average human, this is around 60 degrees. The virtual camera also has a **field of view** (**FOV**) whose value determines the horizontal resolution of the virtual world. It is this value that sets the angle of the projection lines in the perspective view and affects the scale of the world, as seen in *Figure 4.9*. A narrow field of view showing what the camera sees on the left and the camera frustum on the right is shown in *(a)* and a wide field of view with the same camera point of view and frustum view is presented in *(b)*:

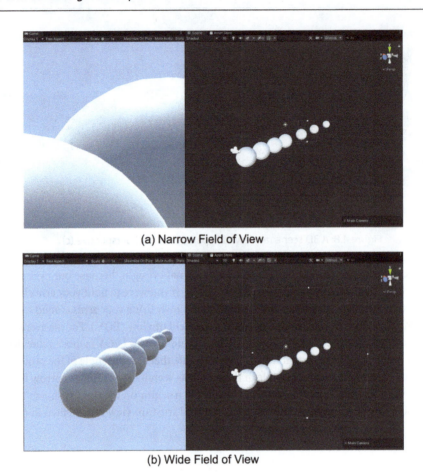

(a) Narrow Field of View

(b) Wide Field of View

Figure 4.9: Different FOVs demonstrated by a camera in the Unity game engine

The FOV for a virtual camera has traditionally been set for the horizontal because of the way humans see the world, though mathematically, there's no reason why the vertical direction couldn't be set; we will stick with the horizontal FOV herein, however. With the horizontal FOV set, the vertical FOV is calculated from the screen resolution using the following equation:

```
vertical_fov = horizontal_fov * screen.height / screen.width)
```

This ensures correct scaling of the environment to ensure the camera's near plane dimensions are a consistent scale of the screen dimensions. Imagine if the camera's near plane were a perfect square and you tried to map this onto a typical 1,920 x 1,080 screen. It would have to stretch the camera's view in the x direction while the y direction remained unaffected.

The mathematical processes that take the vertices of a model and how they are viewed by the camera finally make it into screen coordinates via projections.

Projecting Pixels onto the Screen

As previously discussed, the function of a projection is to map coordinates in the eye space into a rectangular prism with the corner coordinates of (-1, -1, -1) and (1, 1, 1), as shown in *Figure 4.10*. This volume is called **normalized device coordinates** (**NDCs**). This cube is used to define coordinates in a screen-independent display coordinate system and defines the view volume as normalized points. This information can then be used to produce pixels on the screen with the z coordinates being used during the drawing to determine which pixels should appear in front of others. First, though, you need to define the confines of the camera's view volume so that a mapping can be produced:

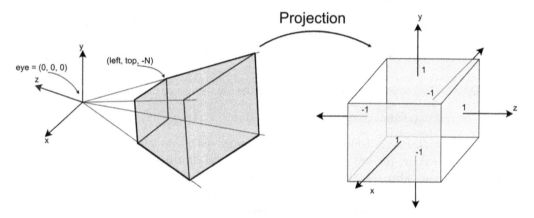

Figure 4.10: The process of projection

Since the object is projected onto the near plane of the camera, we can use the corner coordinates of that plane to determine the mapping. The coordinates of the left top corner are indicated by *(left, top, -N)* where N is the distance of the near plane from the camera's eye. Note that N in this case is negative as the positive direction of the z axis is toward the viewer. Why? To understand this, we must take a brief tangent in our discussion to examine coordinate systems.

Understanding 3D Coordinate Systems in OpenGL

There are two alignments in which 3D coordinates can be defined: the **left-handed system** and the **right-handed system**. Each is used to define the direction in which rotations around the axes occur and can be visualized using the thumb and first two fingers on the respective hands (see *Figure 4.11*). For both systems, the thumb is used to represent the z axis; the direction in which the fingers wrap into a fist is the direction of a positive rotation:

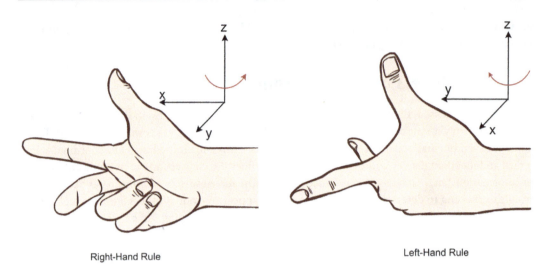

Right-Hand Rule   Left-Hand Rule

Figure 4.11: The right-hand and left-hand rules of 3D axis orientation

For a detailed elucidation, you are encouraged to read the following web page:

`https://en.wikipedia.org/wiki/Right-hand_rule`

In **OpenGL**, eye coordinates are defined using the right-handed system, meaning that the positive z axis protrudes out of the screen (toward the eye) whereas traditionally in **DirectX**, the z axis is positive in the opposite direction. Other 3D software packages seem to have randomly chosen which coordinate system to use. For example, **Blender** adheres to the right-hand rule, and Unity3D is left-handed. The reason for the use of differing rules is simply that no standard exists to define what should be used.

Perspective Projection

A **perspective projection** is how our human eyes view the world, whereby objects get smaller the further they are away from us. We will consider the mathematics of a perspective projection for projecting a point from eye space into NDC using the right-handed system (simply because I'm more familiar with OpenGL than DirectX!). This means eye coordinates are defined with the positive z axis coming out of the screen and placing the camera's near plane z coordinate at $-N$.

To work out the value of left, we make a right-angled triangle between the eye and the middle of the near plane, as shown in *Figure 4.12*. Where the right angle occurs has an x value of 0, with left being in the negative x direction and right in the positive x direction:

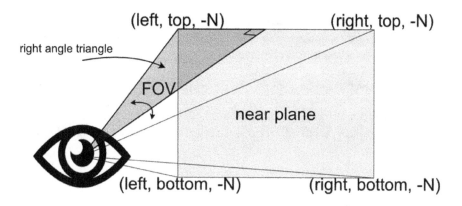

Figure 4.12: The coordinates of the camera's near plane and constructing a right-angle triangle

As the value of *N* is known, using the rules of trigonometry (see *Chapter 8, Reviewing Our Knowledge of Triangles*), we can calculate the following:

```
left = -tan(horizontal_fov / 2) * -N
right = tan(horizontal_fov / 2) * -N
```

`left` can further be defined as follows:

```
left = -right
```

Note that we divide the FOV by 2 as the right-angle triangle only uses half the angle.

Next, to calculate the values of top and bottom, we can do the same thing except with `vertical_fov`, like so:

```
top = tan(vertical_fov/2) * -N
bottom = -tan(vertical_fov/2) * -N
```

At this point, we have the coordinates for all the corners of the near plane. From here, we can calculate the following:

```
nearplane_width = right - left
nearplane_height = top - bottom
```

Let's look at an example of working with these values.

To calculate the `left` and `top` values of a camera, as shown in *Figure 4.12*, with a `horizontal_fov` value of 60, a near plane of 0.03, and a screen resolution of 1920 x 1080, first we must work out the `vertical_fov` value. Recall this is obtained like so:

```
vertical_fov = horizontal_fov * screen.height / screen.width)
```

For this camera, that makes `vertical_fov` equal to 33.75. We can then calculate `left` as -tan(30) x 0.03, which is -0.017, and `top` as tan(33.75/2), which is 0.3. But don't take my word for it. Calculate it yourself.

Hint

When working with mathematics that returns values that can be plotted or illustrated, don't take the values you calculate at face value. You can provide some validation to the numbers you are getting by considering whether they make sense. Draw a diagram and have a look. For example, if a widescreen has a `horizontal_fov` of 60 then, its `vertical_fov` is first going to be smaller, and considering the screen resolution where the height is kind of close to half the width, then a `vertical_fov` of around half the `horizontal_fov` makes sense.

The calculations presented in this section are essential for when you want to develop your own perspective projections. By working logically through the trigonometry involved in defining the near plane of a camera based on the FOV, we have explored how to derive these equations that will later be used to program a camera.

Your turn...

Exercise B: What are the right and bottom values for a camera's near plane corners if the `horizontal_fov` is 80, the near plane is 2, and the screen resolution is 1,366 x 768?

Projecting into the NDC

With the near-plane extents defined, we then project the point from eye space into the NDC. This process is illustrated in the next screenshot and progresses thus:

- First, a point (labeled 1 in *Figure 4.13*) in eye space is projected forward onto the near plane of the camera (labeled 2).

- Then, the point is mapped onto the near plane of the NDC (labeled 3). This process uses the ratio of the point's location to the width and height of the near plane to place it in the same relative location on the front plane of the NDC.

- The point is then given a depth value to push it back into the NDC (labeled 4):

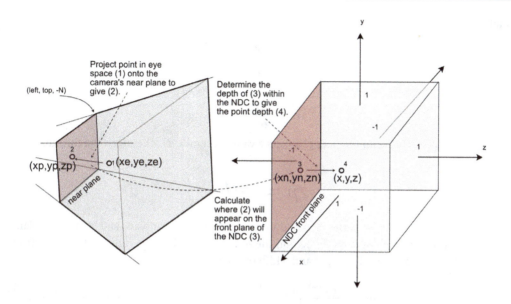

Figure 4.13: The projection process – the steps involved in taking a point from the eye space into the NDC

The point actually stays on the front face of the NDC as this is eventually mapped to the screen coordinates of the device through which the point is being viewed. The depth information is stored purely to determine which points are in front of others and thus which points to cull from drawing if they are behind others and can't be seen.

Points are first projected onto the near plane of the camera where *(xe, ye, ze)* becomes *(xp, yp, zp)*, as illustrated in *Figure 4.13*. We can work out this projection for the *x* coordinates and *y* coordinates separately. This is a straightforward process, using the ratios of similar triangles illustrated in *Figure 4.14*.

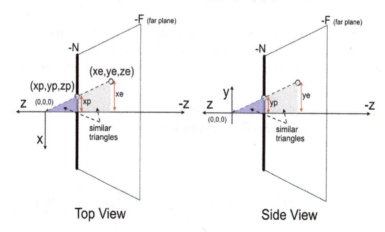

Figure 4.14: Top and side views demonstrating the projection of the x and y values of a point

xp and *yp* can be calculated thus:

$$xp = \frac{-N}{ze} \times xe$$

$$yp = \frac{-N}{ze} \times ye$$

When a point is projected onto the near plane, its *z* value becomes -*N*. Therefore, the formula looks like this:

$$zp = -N$$

The illustration in *Figure 4.15* showing the *x* and *y* dimensions of the near plane and NDC front plane. The location of the same point is on the left on the near plane and on the right on the NDC front plane. Each illustration shows the height and width of each plane:

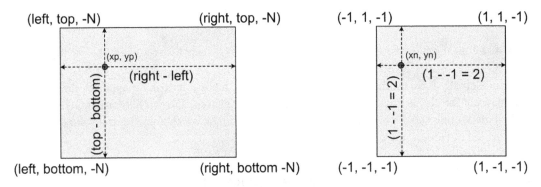

Figure 4.15: A front view of the perspective near plane and NDC front plane

Next, the point *(xp, yp, zp)* is projected onto the front plane of the NDC to produce *(xn, yn, zn)*, as shown in *Figure 4.15*. Solving this problem involves remapping each of the *x*, *y*, and *z* values as a ratio to the respective size of the camera space in each dimension and then determining what this ratio equates to in the NDC.

When the remapping from one plane to another occurs, the ratio of the point to the size of its plane should remain constant. This means that the following formula applies:

$$\frac{xp}{(right - left)} = \frac{xn}{2}$$

$$\frac{yp}{(top - bottom)} = \frac{yn}{2}$$

Put simply, if the value of *xp* were 20% of the way width-wise across the camera's near plane, then the value of *xn* would be 20% across the width of the NDC.

Rearranging to solve for *xn* and *yn*, we get the following:

$$xn = \frac{2 \times xp}{(right - left)}$$

$$yn = \frac{2 \times yp}{(top - bottom)}$$

The value for *zn* can be calculated in the same manner, but instead of using the height and width of a space, the distance in the depth of the space is used thus:

$$\frac{zp}{(N - F)} = \frac{zn}{2}$$

And through rearranging, we get the following:

$$zn = \frac{2 \times zp}{(N - F)}$$

OpenGL takes care of all this mathematics behind the scenes when drawing vertices and figures to the screen. In the next exercise, we will examine perspective and orthographic cameras.

Let's do it...

Through this exercise, you will explore the differing types of camera viewing volume shapes by viewing the cube in each. Follow these next steps:

1. If you haven't already, ensure you have adjusted your `HelloMesh.py` program to draw a cube, as per *Exercise A*. This assumes you have completed the exercise and checked your code against the answers at the end of the chapter.

    Then, add the following line to `HelloMesh.py`:

    ```
    from OpengGL.GLU import *

    . .

    done = False
    ```

```
white = pygame.Color(255, 255, 255)
gluPerspective(30, (screen_width / screen_height),
               0.1, 100.0)
mesh = Cube()
while not done:
    ..
```

`gluPerspective()` allows you to set the camera's FOV, which in this case is 30, then the aspect ratio, followed by the near plane and far plane.

2. Run the program. You'll get a diagonal line drawn across the window, as shown in the following screenshot. So, what happened to the square? It's there—it's just not drawing as you might expect:

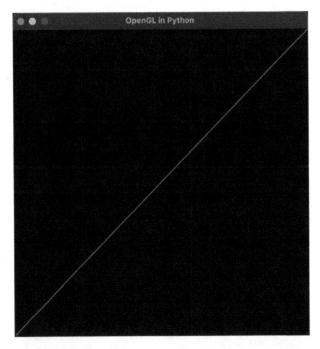

Figure 4.16: The first view from a perspective camera

The reason you are seeing this line across the window is that the cube is being drawn around (0, 0, 0) and the camera is viewing it from (0,0,0). This means the camera is inside the cube and you can only see one side.

3. To get a better view of the cube, the cube needs to be moved into the world along the z axis, like this:

```
    ..
white = pygame.Color(255, 255, 255)
```

```
gluPerspective(30, (screen_width / screen_height),
               0.1, 100.0)
glTranslatef(0.0, 0.0, -3)
mesh = Cube()
  . .
```

Add this line of code and then run HelloMesh again. This time, you will see the entire cube in perspective, as shown in the following screenshot. It doesn't look like much, but the perspective view allows it to be drawn such that the face of the cube furthest away from the camera appears smaller. glTranslate() adds (0, 0, -3) to each of the cube's vertices, thus drawing them at a location further away from the camera:

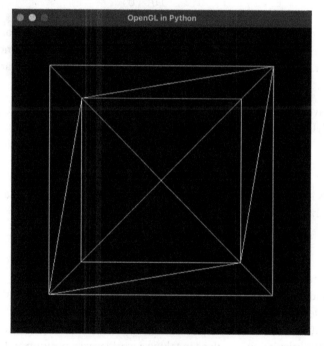

Figure 4.17: A perspective view of a wireframe cube

4. To get a better sense of the 3D nature of the cube, let's rotate it. Add these lines and then run the application again:

```
while not done:
  for event in pygame.event.get():
    . .

  glClear(GL_COLOR_BUFFER_BIT | GL_DEPTH_BUFFER_BIT)
  glRotatef(5, 1, 0, 1)
```

```
mesh.draw()
pygame.display.flip()
pygame.time.wait(100);
```

You'll now see the cube rotating by 5 around its *x* axis and 1 around its *y* axis every 100 milliseconds. At this point, we won't go into the mechanics of these `glRotatef()` and `glTranslatef()` methods as they will be fully explored in later chapters. They are added at this time to help you visualize your drawing better.

5. To change the camera to an orthographic projection, replace the `gluPerspective()` function with the following:

```
glOrtho(-1, 1, 1, -1, 0.1, 100.0)
```

This call is setting, in order, the left, right, top, and bottom near plane and far plane of the view volume. Note that as the cube rotates when it is perfectly aligned with the camera, the planes line up and for a moment only appear as a square. This is because, in orthographic views, objects are not scaled with distance.

In this section, we have stepped through the mathematics behind graphics projection to transform the vertices of a model into screen coordinates for pixels.

Your turn...

Exercise C: Set up a perspective camera to view the cube with a FOV of 60 with the other parameters remaining as previously used. What do you notice about the drawing between FOV = 30 and FOV = 60?

Summary

In this chapter, we've taken a bit of time to focus on how rendering is performed and examined the rendering pipeline and camera setups. A thorough understanding of how these affect the positions and projections of objects in the environment is critical to your understanding of the structure of a 3D environment and the relative locations of drawn artifacts. In addition, considerable time has been spent on elucidating the calculations involved in taking a world vertex position and projecting it onto the screen. The process is not as straightforward as it might first seem, but with a knowledge of basic trigonometry and mathematics, the formulae have been derived.

One common issue that I find when developing these types of applications is just knowing how to debug the code when it seems to be running without syntax errors but nothing appears on the screen. Most often, the issue is that the camera cannot see the object. Therefore, if you can visualize in your mind where the camera is, in which direction it is facing, and whether or not an object is inside the viewing volume, this is a great skill to have.

Another element that assists with visualization in computer graphics is color. This is used from the most basic of drawings, even if they are in black and white, to complex textured surfaces and lighting effects. In the next chapter, we will take a look at the fundamentals of color and how to apply it in computer graphics through code.

Answers

Exercise A:

This is one set of vertices and triangles to draw a cube placed in a new class (note that you might have them in a different order from what is shown here):

```
from Mesh3D import *
class Cube(Mesh3D):
def __init__(self):
            self.vertices = [(0.5, -0.5, 0.5),
(-0.5, -0.5, 0.5), (0.5, 0.5, 0.5),
(-0.5, 0.5, 0.5),(0.5, 0.5, -0.5),
(-0.5, 0.5, -0.5), (0.5, -0.5, -0.5), (-0.5, -0.5, -0.5),
(0.5, 0.5, 0.5), (-0.5, 0.5, 0.5), (0.5, 0.5, -0.5),
(-0.5, 0.5, -0.5), (0.5, -0.5, -0.5), (0.5, -0.5, 0.5),
(-0.5, -0.5, 0.5), (-0.5, -0.5, -0.5), (-0.5, -0.5, 0.5),
(-0.5, 0.5, 0.5), (-0.5, 0.5, -0.5), (-0.5, -0.5, -0.5),
(0.5, -0.5, -0.5), (0.5, 0.5, -0.5), (0.5, 0.5, 0.5),
(0.5, -0.5, 0.5)
]
            self.triangles = [0, 2, 3, 0, 3, 1, 8, 4, 5, 8,
  5, 9, 10, 6,
  7, 10,
  7, 11, 12, 13, 14,
  12, 14, 15, 16, 17, 18, 16,
  18, 19, 20, 21, 22, 20, 22, 23]
```

To use this class in `HelloMesh.py`, include `from Cube import *` and replace `mesh = Mesh3D()` with `mesh = Cube()`.

Because of the point of view, the drawing will look the same as the single square because the other sides of the cube are on top of each other.

Exercise B:

Given `horizontal_fov` = 80, N = 2, `screen.width` = 1366, and `screen.height` = 768, this is the result:

```
vertical_fov = 80 * screen.height/screen.width = 45
right = tan(80/2) * 2 = 1.68
bottom = -tan(45/2) * 2 = -0.83
```

Exercise C:

The greater the FOV, the more the object is scaled, as illustrated here:

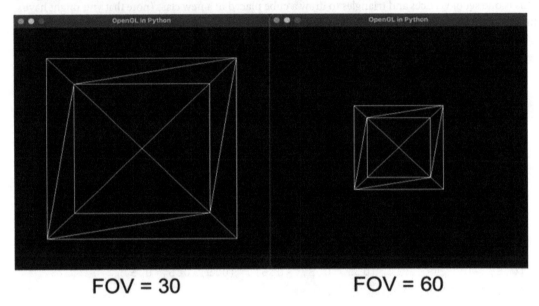

Figure 4.18: A wireframe cube presented with differing FOVs

5
Let's Light It Up!

While we will take a closer look at rendering techniques in depth later on in the book, it always brings a sense of achievement when you can get some color on the screen. Not just a sprite and a pretty background, but painting 3D objects with light and texture.

In this chapter, I will go over the fundamental ways that OpenGL works with **light** and **textures** to give you more to experiment with as you continue to investigate these types of projects. You will learn how to add lighting and materials to the project we are developing throughout. It will provide you with the skills to develop better-rendered images that will assist you in visual confirmation that your code is working as it should.

In this chapter, we will cover the fundamentals of visualization, including the following:

- Adding Lighting Effects
- Placing Textures on Meshes

Technical requirements

In this chapter, we will be using Python, PyCharm, and Pygame, as used in previous chapters.

Before you begin coding, create a new folder in the PyCharm project for the contents of this chapter called `Chapter_5`.

The solution files containing the code can be found on GitHub at `https://github.com/ PacktPublishing/Mathematics-for-Game-Programming-and-Computer- Graphics/tree/main/Chapter05`.

Adding Lighting Effects

Lighting in computer graphics serves the same purpose as it does in the real world. It provides definition to the rendering of objects, making them seem embedded in the virtual world. The way that light is perceived depends on the surface treatment of the objects. These treatments are defined as **materials** and will be discussed in the next section.

Light sources consist of three primary lighting interactions:

- **Ambient light** represents the background light in an environment that doesn't emanate from anywhere. It's like the light that you can still see in a room when all the lights are out, but you can still make out the shapes of objects because light might be leaking into the room through windows. It is light that has been reflected and scattered so much that you can't tell where it is coming from. It's a basic coloring and although it doesn't emanate from a particular source, in computer graphics, it is assigned as part of a light source.

- **Diffuse light** comes from the light source and illuminates the geometry of an object, giving it depth and volume. The surfaces of the object that face the light are the brightest.

- **Specular light** bounces off a surface and can be used to create shiny patches. As with diffuse light, specular light comes from the direction of the light source, and the surfaces of the object that face the light are the brightest.

Each of these can have a color assigned to produce differing effects. The results of each can be seen in *Figure 5.1*:

Figure 5.1: The components of light that make up a final render

With the exception of ambient light, which is applied uniformly over a model, the way light is seen on an object is by the way that it is reflected. There are numerous models, called **shading models**, that define the mathematics involved in generating this light on surfaces in graphics. The two types are called **diffuse scattering** and **specular reflection**. While these will be elucidated in later chapters, it is worth an overview of how they work before we try them in Python with OpenGL. *Figure 5.2(a)* illustrates the important directions involved when calculating light intensities. The light source shines at a point on a surface. The light from this point is reflected to the viewer. In the simplest illumination model, the light will be seen at its most intense when viewed at the same reflected angle around the surface normal (indicated by **n**).

Diffuse scattering (see *Figure 5.2(b)*) occurs when incoming light is partially absorbed by a surface and is re-radiated uniformly in all directions. Such light interacts with the surface material of the object and as such, the light's color is affected by the surface color. It makes no difference which angle the surface is viewed from. In *Figure 5.2(b)*, a viewer at **v1** and a viewer at **v2** will experience the same intensity of light coming from the surface:

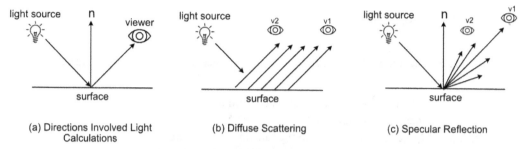

(a) Directions Involved Light   (b) Diffuse Scattering   (c) Specular Reflection
Calculations

Figure 5.2: The components of light that make up a final render

Specular reflections are highly directional. Unlike diffuse scattering, it does matter where the view is in relation to the light source. The intensity of the light will be strongest where the incoming light angle and reflected light angle are the greatest but then drop off away for angles greater or less than. For example, in *Figure 5.2(c)*, a viewer at **v1** will see a stronger light intensity than the viewer at **v2**.

It's probably time to look at turning on the lights in a Python application.

Let's do it...

In this exercise, we will explore how lights are enabled and used in OpenGL. Be sure to create a new folder called `Chapter_Five` in which to place the files for these exercises, and then follow these steps:

1. Create a new Python script called `HelloLights.py` and copy the exact same code from the previous rotating cube in the *Chapter 4, Graphics and Game Engine Components* script in `HelloMesh.py`.

2. Edit `Mesh3D.py` and change the line from this:

    ```
    glBegin(GL_LINE_LOOP)
    ```

    Change it to this:

    ```
    glBegin(GL_POLYGON)
    ```

3. Run this application. You will get a black window with the white silhouette of a solid rotating cube.

4. Now, modify `HelloLights.py` with the following lines:

    ```
    done = False
    white = pygame.Color(255, 255, 255)
    glMatrixMode(GL_PROJECTION)
    gluPerspective(60, (screen_width / screen_height),
    ```

```
                                0.1, 100.0)
glMatrixMode(GL_MODELVIEW)
glTranslatef(0.0, 0.0, -3)
glEnable(GL_DEPTH_TEST)
mesh = Cube()
```

The code has become a little more complex and introduces a few new OpenGL calls, including the following:

- glMatrixMode(): As we have not yet covered matrices in this book, for now, just think of them as changing coordinate spaces. There are two calls for glMatrixMode(). First, we go into the projection view to set up the camera and its coordinate system. Next, we change to the modelview that allows us to specify coordinates in the world space. Working in world space coordinates is far more intuitive.

- glEnable(GL_DEPTH_TEST): Ensures that fragments (pixels yet to be drawn) behind other fragments with respect to the camera view are discarded and not processed, making rendering more efficient.

5. Let's turn on some lights. Add in the code to enable them:

    ```
    ..
    glTranslatef(0, 0, -4)
    glEnable(GL_DEPTH_TEST)
    glEnable(GL_LIGHTING)
    while not done:
    ..
    ```

    Now, when you run the application, the lights will be on, but they might not look like they are because you will see a very dull version of the rotating cube. This is a kind of ambient lighting. Just because the lights are enabled doesn't spontaneously create any lights. We have to do that manually.

6. To create a light and turn it on, add the following:

    ```
    ..
    glEnable(GL_DEPTH_TEST)
    glEnable(GL_LIGHTING)
    glLight(GL_LIGHT0, GL_POSITION, (5, 5, 5, 1))
    glEnable(GL_LIGHT0)
    while not done:
    ..
    ```

Let's look at what we've done here:

- `glLight()` defines the first light called `GL_LIGHT0` as indicated by the first parameter. This is an internal symbol used by OpenGL and as such, you can have `GL_LIGHT1`, `GL_LIGHT2`, and so on, up to `GL_MAX_LIGHTS` (which is an OpenGL internal value you can print out to determine how many lights you have access to with your graphics card).

- The second parameter of `glLight()` indicates what you want to do with the light. In this case, you are setting the position and placing it at `(5, 5, 5)` in world coordinates. You will note here that instead of a three-coordinate value for the light position, it is given as four values. The fourth value is w and is important in matrix calculations. For now, just leave it as `1`.

- `glEnable(GL_LIGHT0)` turns on the light.

7. To control the ambient, diffuse, and specular colors from `GL_LIGHT0`, add the following:

```
..

glLight(GL_LIGHT0, GL_POSITION, (5, 5, 5, 1))
glLightfv(GL_LIGHT0, GL_AMBIENT, (1, 0, 1, 1))
glLightfv(GL_LIGHT0, GL_DIFFUSE, (1, 1, 0, 1))
glLightfv(GL_LIGHT0, GL_SPECULAR, (0, 1, 0, 1))
glEnable(GL_LIGHT0)

..
```

8. Run the application to see a fully lit cube, as shown in *Figure 5.3*:

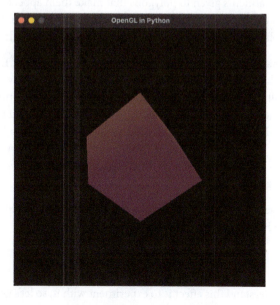

Figure 5.3: A cube with lighting enabled

Each of the given light types is succeeded by a four-valued *red, green, blue, and alpha (transparency)* value. The light will be positioned in the world at (5, 5, 5) and have a magenta ambient color, a yellow diffuse color, and a green specular color. Note that OpenGL requires color channel values to be specified between 0 and 1.

You can run this application to see the effect as well as play around with turning the light types off and changing the color. Be aware that at this stage, you might find it difficult to get any specular lighting to appear with these very few settings. It's sometimes easier to see if you revolve the model around, as there will be a glint in the corners.

Your turn...

Exercise A: Change the ambient color of GL_LIGHT0 to green.

Exercise B: Add a new light symbolized by GL_LIGHT1 and place it at (-5, 5, 0). Set its diffuse light to blue and don't forget to enable the light.

There's so much more to lighting systems in graphics than has been covered here. Although they are very complex, at this stage in the book, it's just exciting to see some lights on the models we are rendering.

In addition to lights, the surface treatment of objects is also important to their appearance. We'll add some materials in the next section.

Placing Textures on Meshes

Materials are the surface treatments given to polygons that make them appear solid. The coloring-in of the polygon plane gives the illusion that it has substance and is more than its surrounding edges. The surface treatment applied interacts with the lighting applied to give the final appearance.

Just like lights, materials have different ambient, diffuse, and specular colors. Each of these interacts with the corresponding lights to determine the final effect seen. For example, white diffuse light shone on a diffuse green cube will reflect the color green. Although white light is hitting the cube, the green color of the cube determines what light gets reflected. White light is when all the color channels for R, G, and B are turned on, hence the (1, 1, 1) value. If the cube is green, then its color is set to (0, 1, 0). Essentially, it's the channel that the light and the material both have turned on that is reflected. Of course, it is a little more complex than that, as color channels can take on any value between 0 and 1, but you get the idea.

The process of adding materials is possibly even more complicated than lighting and there's no better way of learning how to code with them than jumping right in.

Let's do it...

The best way to see and understand this effect is to experiment with it, so let's get started by following these steps:

1. From your previous `HelloLights.py` code, remove any extra lights that you might have created and for the single light left, set only the diffuse color to (1, 1, 1) and add these extra lines, which will add a material onto the cube:

```
glEnable(GL_LIGHTING)
glLight(GL_LIGHT0, GL_POSITION, (5, 5, 5, 1))
glLightfv(GL_LIGHT0, GL_DIFFUSE, (1, 1, 1, 1))
glEnable(GL_LIGHT0)
glMaterialfv(GL_FRONT, GL_DIFFUSE, (0, 1, 0, 1));
while not done:
```

Run this to see the green cube. White light falling onto a green object will display that object as green.

2. Now change the light's color from white to yellow, for example, (1, 1, 0, 1). Run this. The cube is still green. Why? Well, the cube is green, and the light has both red and green channels. The red channel is absorbed by the teapot while the green channel is reflected. So, the effect is the same as before.

3. Now change the light's color from yellow to red, for example, (1, 0, 0, 1). Run this. You get a very dark colored almost black, except for that background ambient light cube. Why? Because the red light shining on the cube is fully absorbed and there's nothing left to reflect.

Besides coloring the surface of an object with lights and pure color, an external image can be used. This image is called a **texture** as it can be used to add a haptic quality to the object's surface. Textures are placed on the surface of an object, polygon by polygon, through a process called **texture mapping**. This mapping occurs by giving each corner of the texture a two-dimensional coordinate with values ranging between 0 and 1. The horizontal values of the texture are akin to the *x* axis; however, in this process, they are referred to as u. The vertical values, like the *y* axis, are referred to as v. Hence the coordinates of a texture are called **UVs**. The UV values are matched to specific vertices as shown in *Figure 5.4*; the UVs that match the bottom-right triangle of a square are highlighted:

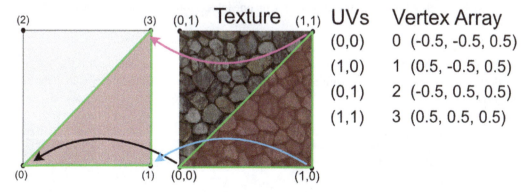

Figure 5.4: Texture Mapping

Working with External Files

From this point onwards in the book, we will be working with files that are external to the Python script. Throughout the book, the code assumes the external files (models, textures and shaders) are in the same folder as the code accessing them. However, in GitHub repository of this book, the external files are accessed as if one folder higher. Please keep this in mind for referencing external files in the code.

For example, if your project files are organized like this:

```
Root  folder
- Project 1
--- Main.py
--- Models
------ Teapot.obj
```

It means that the `models` folder containing the teapot model is in the same folder as the `main.py` code. To reference `Teapot.obj` from inside `Main.py` the address is:

```
Models/Teapot.obj
```

However, if you would like to keep the external files in one place and outside of any particular chapter project, you can place them on the same level as each chapter code (as is referenced in GitHub repository of this book). This structure appears as:

```
Root folder
--- Project 1
------ Main.py
--- Models
------ Teapot.obj
```

In this case, the address of `Teapot.obj` should be:

```
../Models/Teapot.obj
```

The `../` in the file address tells the code to look up one folder from where the code is running from.

The best way to get to understand UVs is to start using them. The way that OpenGL binds the texture to vertices might seem a little long-winded, but you will soon understand the process in order to apply it elsewhere.

Let's do it...

Work through these steps to place textures onto the models by controlling the UVs:

1. Before adding in a texture and performing some UV mapping with our cube, we will make a few modifications to the existing `Mesh3D` and `Cube` classes. First, in the `Mesh3D` class, add the following:

    ```
    class Mesh3D:
      def __init__(self):

        ...

        self.triangles = [0, 2, 3, 0, 3, 1]
        self.draw_type = GL_LINE_LOOP
        self.texture = pygame.image.load()
        self.texID = 0
    ```

    These three lines create extra variables we can use to set how a mesh will be drawn and what, if any, texture to use.

    Next, we modify the `Cube` class to set the values of these new variables:

    ```
    class Cube(Mesh3D):
      def __init__(self, draw_type, filename):
        self.vertices = ...
        self.triangles = ...
        Mesh3D.texture = pygame.image.load(filename)
        Mesh3D.draw_type = draw_type
    ```

    Note that we will be passing through `draw_type` and a filename from `main script`. However, before we do that, we should add a texture file.

2. Visit `textures.com` and find a texture that you would like to map onto the cube. Select one that is 512 x 512. Download the texture and add it to PyCharm in the same manner used in *Chapter 1, Hello Graphics Window: You're On Your Way*. Take note of the path to this image for use in your code.

3. Edit the mesh creation line in `HelloLights.py` to accept the new parameters:

    ```
    mesh = Cube(GL_POLYGON, "images/brick.tif")
    ```

    Note that where I have the path to my image, you should replace it with yours.

4. At this point, you might like to run the code to check you haven't introduced any syntax issues. There will be no difference in the cube. It will draw as a solid and use the lighting settings from the previous exercise.

5. Now, add the new method to the `Mesh3D` class:

```
def init_texture(self):
  self.texID = glGenTextures(1)
  textureData = pygame.image.tostring(self.texture,
                                      "RGB", 1)
  width = self.texture.get_width()
  height = self.texture.get_height()
  glBindTexture(GL_TEXTURE_2D, self.texID)
  glTexParameteri(GL_TEXTURE_2D,
                  GL_TEXTURE_MIN_FILTER,
                  GL_LINEAR)
  glTexImage2D(GL_TEXTURE_2D, 0, 3, width, height, 0,
               GL_RGB, GL_UNSIGNED_BYTE, textureData)
```

This method takes the image loaded with Pygame and loads it into OpenGL by doing the following:

- Using `glGenTextures()` to create a unique ID for the texture that can be used by OpenGL.

- Calling `glBindTexture()`, which sets the new ID to the type of texture being created, which in this case is a normal 2D image.

- Running `glTexParameteri()` sets up the function used by the texture for a variety of methods. In this case for working with this `GL_TEXTURE_2D`, the function of `GL_TEXTURE_MIN_FILTER` (which determines how lower resolutions of a texture are used to blur an image with distance from the camera) is set to `GL_LINEAR`, which provides a linear distance falloff for this blurring effect. This is also referred to as **mipmapping** and will be covered in depth in later chapters.

- Generating the OpenGL texture from the given image file and binding it to the texture ID given in the preceding `glBindTexture()` call.

6. The preceding method can now be called at the very bottom of the `__init__()` method in the `Cube` class:

```
Mesh3D.texture = pygame.image.load(filename)
Mesh3D.draw_type = draw_type
Mesh3D.init_texture(self)
```

7. With all this setup done, it's time to define the UV values. There needs to be one UV for each vertex that has been given where the UVs are in the same order as the vertex to which they are mapped. In the Cube class, add this new set of UVs like this:

```
self.triangles = …
self.uvs = [(0.0, 0.0),
       (1.0, 0.0),
       (0.0, 1.0),
       (1.0, 1.0),
       (0.0, 1.0),
       (1.0, 1.0),
       (0.0, 1.0),
       (1.0, 1.0),
       (0.0, 0.0),
       (1.0, 0.0),
       (0.0, 0.0),
       (1.0, 0.0),
       (0.0, 0.0),
       (0.0, 1.0),
       (1.0, 1.0),
       (1.0, 0.0),
       (0.0, 0.0),
       (0.0, 1.0),
       (1.0, 1.0),
       (1.0, 0.0),
       (0.0, 0.0),
       (0.0, 1.0),
       (1.0, 1.0),
       (1.0, 0.0)]
Mesh3D.texture = pygame.image.load(filename)
```

8. With the UV values in place, the draw() method in the Mesh3D class can be updated:

```
def draw(self):
  glEnable(GL_TEXTURE_2D)
  glTexEnvf(GL_TEXTURE_ENV, GL_TEXTURE_ENV_MODE,
            GL_DECAL)
  glBindTexture(GL_TEXTURE_2D, self.texID)
```

```
for t in range(0, len(self.triangles), 3):
  glBegin(self.draw_type)
  glTexCoord2fv(self.uvs[self.triangles[t]])
  glVertex3fv(self.vertices[self.triangles[t]])
  glTexCoord2fv(self.uvs[self.triangles[t + 1]])
  glVertex3fv(self.vertices[self.triangles[t + 1]])
  glTexCoord2fv(self.uvs[self.triangles[t + 2]])
  glVertex3fv(self.vertices[self.triangles[t + 2]])
  glEnd()
```

OpenGL draws through the enabling and disabling of facilities with differing method calls. These switch OpenGL into different processing states. In this case, at the beginning of the draw() method, texture mapping is enabled with glEnable(GL_TEXTURE_2D). The glTexEnvf() function sets up the texture environment that determines how texture values are interpreted for the purpose of generating an *RGBA* color value. Put simply, the GL_DECAL parameter keeps the incoming texture pixel *RGBA* values the same as they are in the source image. The particulars of this method are quite complex and for more information, you are encouraged to look at the following: https://www.khronos.org/registry/OpenGL-Refpages/gl2.1/xhtml/glTexEnv.xml.

Next, you will notice the use of glBindTexture() again. This tells OpenGL which texture we want to work with.

Following this, the code enters the for loop previously used to draw the wireframe and solid cubes. Notice the code remains relatively the same with the exception that the UV values are set before the vertices they match with. As we have a UV value for each vertex and each triangle keeps track of the vertices used, the code can also refer to the correct UV values.

9. Run the code. You will see a rotating solid cube with the texture given mapped across the surface, as shown in *Figure 5.5*:

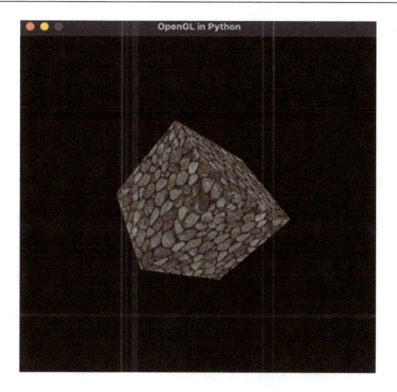

Figure 5.5: A texture mapped onto a cube

This exercise has covered UV values and their use for mapping a texture onto a 3D object. As you can see, the process involves a lot of arrays of values. These arrays include the vertex list and another list for UVs where each vertex is assigned a UV in order to get the texture perfectly aligned with the geometry.

Summary

This chapter has been a short examination of lighting and texture that help to bring depth to a scene and make objects appear solid. Besides the lists of vertices and UVs that we've used, you will discover later in the book that a 3D mesh can possess many other sets of values used for a multitude of rendering effects. But for now, you have enough skill in displaying a simple mesh and adding a texture to it.

In the next chapter, we will add more flexibility to the project we are creating by working more with the `main` loop. In addition, we will improve the application architecture to make it more extensible by giving simple rendered objects access to more behaviors.

Answers

Exercise A:

glLightfv(GL_LIGHT0, GL_AMBIENT, (0, 1, 0, 1))

Exercise B:

```
glLightfv(GL_LIGHT0, GL_SPECULAR, (0, 1, 0, 1))
glEnable(GL_LIGHT0)
glLight(GL_LIGHT1, GL_POSITION, (-5, 5, 0, 1))
glLightfv(GL_LIGHT1, GL_DIFFUSE, (0, 0, 1, 1))
glEnable(GL_LIGHT1)
while not done:
```

6
Updating and Drawing the Graphics Environment

All objects in a graphics environment undergo a cyclical process of being updated and then drawn. This occurs in the main game loop and is synchronized by a frame rate clock. Building this ability into our graphics engine and application at this point is critical for further functionality, such as physics and other interactivity, to be added down the line. Adding a strong foundation to facilitate this early on is crucial in order to succeed in graphics processing and rendering down the line.

Herein, we will examine the purpose of the main game loop and add a clock to regulate the frame rate in our project. Coordinated with this loop are updates to objects so that they can become whatever we need them to be, from audio sources to game characters. Therefore, we will also be concentrating our efforts on developing an object's abilities.

To this end, in this chapter, we will be examining the following:

- Introducing the main game loop
- Updating and drawing objects
- Measuring time

By the end of this chapter, you will have gained an understanding of how objects in a game engine can take on individual properties that allow them to serve a variety of purposes. This will include developing a code architecture to support this functionality.

Technical requirements

In this chapter, we will be using Python, PyCharm, and Pygame as in the previous chapters. Before you begin coding, create a new folder called `Chapter 6` in the PyCharm project for the contents of this chapter.

The solution files containing the code can be found on GitHub at `https://github.com/PacktPublishing/Mathematics-for-Game-Programming-and-Computer-Graphics/tree/main/Chapter06` in the `Chapter06` folder.

Introducing the main game loop

To keep any windowed application alive, the program must instigate an endless loop. We created one in *Chapter 1*, *Hello Graphics Window: You're On Your Way*, to keep the graphics window open in the first exercise through the implementation of a `while` loop. Since then, we've continued to use this design pattern in every application. This loop in a game is called the **main game loop**. Each loop produces one frame of graphics. In Pygame, this frame is pushed to the screen using the `pygame.display.update()` or `pygame.display.flip()` methods.

Thus far, our graphics applications have been very simple with very little happening within the loop – however, in a full-fledged game engine, the loop is quite complex and contains multiple significant events, as illustrated in *Figure 6.1*:

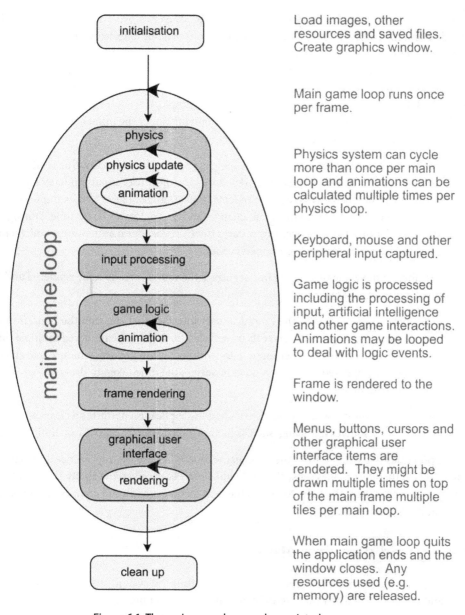

initialisation

Load images, other resources and saved files. Create graphics window.

Main game loop runs once per frame.

physics

physics update

animation

Physics system can cycle more than once per main loop and animations can be calculated multiple times per physics loop.

main game loop

input processing

Keyboard, mouse and other peripheral input captured.

game logic

animation

Game logic is processed including the processing of input, artificial intelligence and other game interactions. Animations may be looped to deal with logic events.

frame rendering

Frame is rendered to the window.

graphical user interface

rendering

Menus, buttons, cursors and other graphical user interface items are rendered. They might be drawn multiple times on top of the main frame multiple tiles per main loop.

clean up

When main game loop quits the application ends and the window closes. Any resources used (e.g. memory) are released.

Figure 6.1: The main game loop and associated processes

After the application begins running, various items are initialized and loaded, such as images, sounds, and other resources, and the main graphics window is created, the program enters an endless loop. The loop continues to run until the user quits the application. The main game loop contains various subsections, which we are yet to explore through code – however, we will briefly examine them now.

The physics system is a major part of most games, more so in 3D games. Most of the work undertaken by it is for determining collisions and calculating the reactions of game objects. As the physics applied to one game object may affect other game objects, such as two balls colliding, the physics system maintains its own loop, and this may need to loop several times per main game loop to ensure the positions of all game objects have been updated before they are rendered. Within the physics system, procedural animations can be updated many times. For example, if we were using code to place a character's hand on a moving object and that object was moving under the influence of physics, then the animations would need to be updated as the object moved.

After the user input has been processed, the game logic is processed. Essentially, this is all the code you've written that determines how the game works. It includes how user input influences the game environment, what should occur when certain objects impact other objects (for example, a bullet hits a character), and how the game environment changes and moves forward over time. During this process, another animation loop may occur to process further procedural animations, which could be influenced by user input or other programmed events.

Following this, after all the game processing has occurred, the frame is ready to be rendered and the pixels are drawn in the graphics window.

Before the main loop finishes its cycle, the graphical user interface that sits atop the main frame is rendered. In some game engines, this can have its own cycle, causing it to be rendered multiple times per main graphics loop. This allows for user input in the form of button presses or mouse movements, among other things, to be updated and drawn quickly despite the time taken to draw a single frame of the main game environment.

After the main loop ends, the graphics window closes and the application performs a clean-up process that releases any system resources it is using, such as memory allocation, back to the operating system.

Many of the processes that occur within the main game loop influence the appearance and behavior of drawn objects. It therefore makes sense that individual objects are given the ability to control their own updates within each loop. To facilitate this, in the next section, we will add components to objects that will embody this functionality.

Updating and drawing objects

Objects in a game or graphics application are entities within the system that possess numerous properties and methods that allow them to exist and interact within their environment.

As shown in *Figure 6.2*, a typical object possesses many subparts or components:

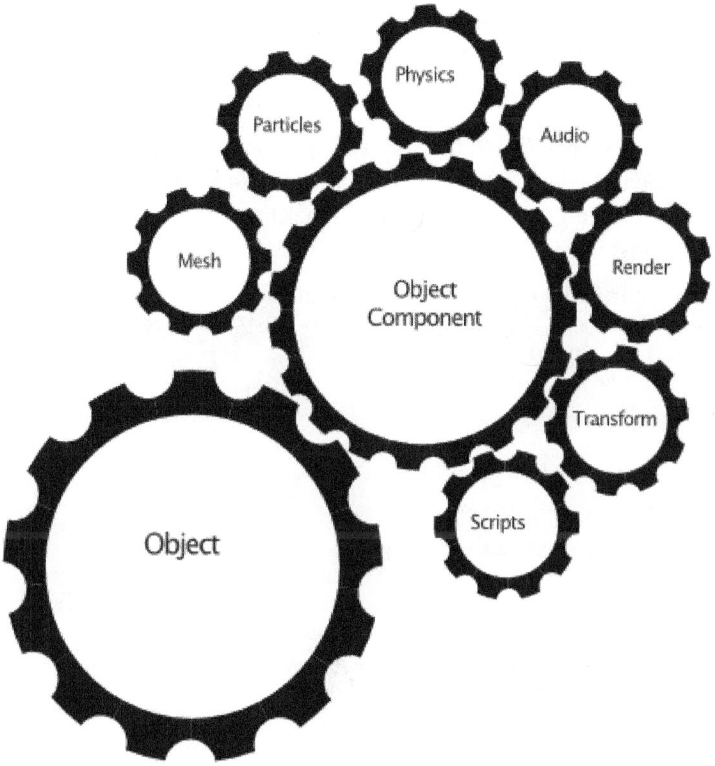

Figure 6.2: A game object and a few of its components

You can think of components as functionality added to the object, for without the components, the object is nothing except a placeholder. The object may have all or only a subset of the components, as illustrated in *Figure 6.2*. For example, you might have an object in the environment that only plays a sound. For this, it would only need an audio component, whereas an object representing a game character might have a mesh that defines what it looks like, a render component that tells the game engine how to draw it on the screen, and a transform component that stores its location and orientation in the world.

In a way, you might consider the cube that we created in the exercises in *Chapter 4, Graphics and Game Engine Components*, as a component. However, our code isn't structured in the same way that it is in *Figure 6.2*, so it's time to fix that.

As we continue working on the Python/OpenGL project through the book we will start making classes inspired by each of the components in *Figure 6.2*. We will start by creating an object class.

Let's do it...

In this exercise, we will start turning our graphics project into a real engine through the addition of components. Don't forget to create a new folder for the code in this chapter. Let's get started with the exercise:

1. Create a new Python script called `HelloObject.py`. Copy the code from the `HelloLights.py` you completed in *Chapter 5, Let's Light It Up!*, or grab a fresh copy of the one included with `Chapter 5` on GitHub. You will also need to make copies of `Cube.py` and `Mesh3D.py` in the new `Chapter 6` folder.

2. Create a new Python script called `Object.py` and add the following code:

```python
from Mesh3D import *


class Object:
    def __init__(self, obj_name):
        self.name = obj_name
        self.components = []


    def add_component(self, component):
        self.components.append(component)


    def update(self):
        for c in self.components:
            if isinstance(c, Mesh3D):
                c.draw()
```

In this, we have created a generic object class that is capable of storing multiple components. Components are added with the `add_component()` method into a list which is then searched through to find components constructed with the Mesh3D class in `update()`.

3. Be sure to include `Object` at the top of `HelloObject.py` as in the following:

```python
from Object import *
```

4. Modify the code in `HelloObject.py` as follows to make use of the new `Object` class:

```python
. .
glTranslatef(0.0, 0.0, -3)
glEnable(GL_DEPTH_TEST)
cube = Object("Cube")
cube.add_component(Cube(GL_POLYGON,
```

```
                          "images/wall.tif"))
    glEnable(GL_LIGHTING)
    ..
```

Note that you must remove the previous variable, mesh, and replace it with the instantiation of a cube as an object and then add the cube mesh component to it. Within the main loop, be sure to replace mesh.draw() with cube.update().

5. Run the HelloObject.py script. You will not notice any changes in its current functioning.

6. We will now add another component to create a transform. Create a new script called Transform.py and add the following code:

```
import pygame


class Transform:
    def __init__(self, position):
        self.set_position(position)

    def get_position(self):
        return self.position

    def set_position(self, position):
        self.position = pygame.math.Vector3(position)
```

This component defines the object's location in 3D space. Notice the use of the Pygame Vector3 class. You will be seeing this a lot. Besides other things, this class stores a tuple in the form of x, y, and z. This makes the code more intuitive to read.

7. To integrate the new Transform class, we need to add it as a component in the HelloObject.py script first as follows:

```
from Mesh3D import *
from Transform import *
..
class Object:
..
glEnable(GL_DEPTH_TEST)
cube = Object("Cube")
cube.add_component(Transform((0, 0, -1)))
cube.add_component(Cube(GL_POLYGON,
                        "images/wall.tif"))
```

By including the cube.add_component(Transform((0, 0, -1))) line, you are setting the location of the cube to (0, 0, -1) – however, you must remember that earlier in the code from *step 4* we set the following:

```
glTranslatef(0.0, 0.0, -3)
```

This is already offsetting whatever is drawn by z = -3. Therefore, by adding the Transform component with z = -1, the total offset for the cube will be the addition of these, which equates to z = -4.

8. To get the Transform component to affect the drawing, further changes need to be made to the add_component and update functions in the Object class. Therefore, next we test if the component is of type Transform and push it into first position in the list.

```
from Transform import *

..

def add_component(self, component):
    if isinstance(component, Transform):
        self.components.insert(0, self.components)


    self.components.append(component)


def update(self):
    for c in self.components:
        if isinstance(c, Transform):
            pos = c.get_position()
            glTranslatef(pos.x, pos.y, pos.z)
        if isinstance(c, Mesh3D):
            c.draw()
```

There are a couple of things to take note of in this code. First, a Transform component is treated as a very special case, and you will see that it is inserted first in the component list. This is because the position needs to be set before any drawing is done by Mesh3D. Next, the update method, before drawing the mesh, calls a glTranslatef() method to ensure that the movement from the Transform component is taken into consideration and sent to OpenGL.

9. It's time to test the program out again by running it. Before you do, try and imagine what you might see. I can bet you won't see what's about to come. Run your code (and get it running) before moving on to the next point.

If your cube flew off into the distance, as shown in *Figure 6.3*, then the program is running as it should:

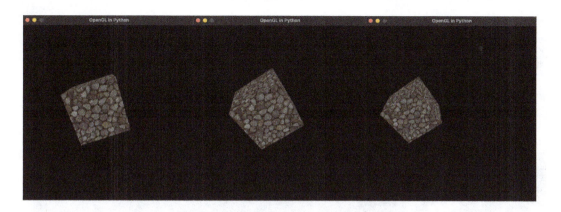

Figure 6.3: A rotating cube moving off into the distance

Here's what's happening. In each main loop, the `Object` class's `update` method is called. This runs a `glTranslatef()` method before drawing the mesh. It does this for every frame. The thing with `glTranslatef()` is that it's compounding. So, in each loop, z = -1 is added to the position of the cube and this animates it flying off into the distance. So, how do we fix this?

10. OpenGL has a very clever mechanism. It's able to store the current state of the environment, draw something, and then return to the previous state. What we want to do is basically save the position of the first `glTranslate(0, 0, -3)`, add the `glTranslate(0, 0, -1)` method, draw the cube, and then take away the last `glTranslate()`. We do it by modifying the `update` method in `Object` to the following:

```
def update(self):
    glPushMatrix()
    for c in self.components:
        if isinstance(c, Transform):
            pos = c.get_position()
            glTranslatef(pos.x, pos.y, pos.z)
        if isinstance(c, Mesh3D):
            c.draw()
    glPopMatrix()
```

In short, think of `glPushMatrix()` as fixing the graphics environment with respect to transformations (including a translate). This causes OpenGL to remember the current state of the environment. When `glPopMatrix()` is called, OpenGL returns to the last remembered state of the environment. Don't worry too much about the technical details that are going on in the background at this time, as we will cover them in detail in later chapters.

For now, though, you can run your code to check that the cube doesn't fly off into the distance anymore, as illustrated in *Figure 6.4*.

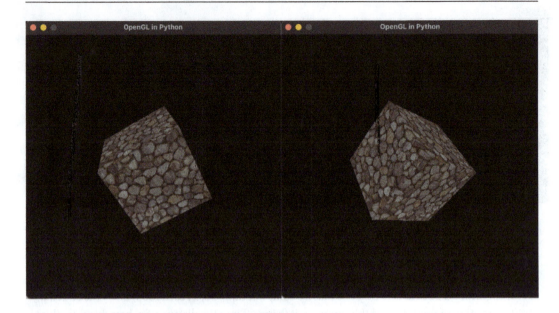

Figure 6.4: A cube rotating in place

Its rotation will look slightly different and this is because we've added the extra translate inside the object's `update` method. If you set this translate to `(0, 0, 0)`, the rotation animation will look exactly as it did before.

Your turn...

Exercise A: Add another cube to the scene that has a different texture and a translate of `(0, 1, 0)`. To ensure all objects in the scene have their `update` method called, create a list called `objects = []` before your cube creation code. Duplicate the three lines of code you used to create the original cube, change `cube` to `cube2`, and append each new object into the list. Loop through the objects list in the main loop to call the update methods of each, not forgetting to remove the original call to update.

In this section, we've seen how to update and draw multiple graphics objects. To ensure consistent frame rates in an application, controlling the speed of the main game loop is crucial. This can be done by setting the clock speed in Pygame.

Measuring time

Time is used in graphics to control animations – these are any moving objects or changing images. It's also useful for determining the frame rate at which your application is running. There are numerous useful functions for this in the Pygame API, and you can read the details here: `https://www.pygame.org/docs/ref/time.html`. The frame rate is expressed in **frames per second (FPS)**, and this is an important measurement as it specifies how fast the main loop takes to execute. It's important to keep the loop running at the same rate for the entire application to ensure smooth performance and animations. If the loop changes speed, then the experience becomes jittery.

Currently, in our code, the frame rate will be as fast as your computer can run it. However, it might not be consistent, as your operating system processes other things in the background. To see just how fast your computer is running the main loop, try removing the following from the code:

```
pygame.time.wait(100)
```

The animated rotation will now go at breakneck speed. Replace it with this instead:

```
print('tick={}, fps={}'.format(clock.tick(),
    clock.get_fps())).
```

Here, the clock is defined above the `while` loop as `clock = pygame.time.Clock()`.

When run, this will print out the tick, which is the number of milliseconds that have passed since the last time `tick()` was called, or the number of milliseconds the last main loop took to execute and the FPS. The type of output you will get in the console may look similar to the following:

```
tick=2, fps=714.2857055664062
tick=1, fps=714.2857055664062
tick=1, fps=714.2857055664062
tick=1, fps=666.6666870117188
tick=2, fps=666.6666870117188
```

Note, `get_fps()` averages the FPS every 10 frames, so you won't get any values initially.

To control this and smooth out the FPS, we can make a simple modification to the code using `pygame.time.Clock()`.

Let's do it...

Add a clock into the main game loop of `HelloObject.py` as follows:

```
clock = pygame.time.Clock()
fps = 30
while not done:
    ..
    pygame.display.flip()
    clock.tick(fps)
pygame.quit()
```

These simple additions will keep the FPS as close as possible to the value set in the `fps` variable. It works by keeping track of how long the previous frame took and then adding a small waiting time where necessary to slow the loop down. Try running this to see how running at 30 FPS slows down the animation compared to not controlling the main loop speed. If you increase the value of `fps`, it will speed up the animation.

Thorough control over the speed of the frame rate and hence, the main game loop is an important feature in a graphics engine. In this section, we've used the clock speed in Pygame to control this to ensure a scene is drawn smoothly.

Summary

In this chapter, the fundamental elements of a graphics rendering engine were explored. It was a whirlwind introduction to numerous concepts that will be thrashed out in later chapters but was essential to gaining an overview of how to move our coding structure forward.

We began by looking at the complexities of the main game loop and the many processes an object moves through to be updated and drawn. Using this knowledge, we then added component functionality to the objects in our project to ready them for the expansion of the engine we are developing. In addition, we also explored how time is an essential controlling component of the main loop and added this practicality to the code base. There was more programming in this chapter than mathematics, but again, a lot of it was laying down a good base on which to build our mathematics functionality.

In the next chapter, we will add functionality to our application that will cater to user input through the integration of methods to manage commands from the keyboard and mouse. This will allow us to program functions to deal with input, such as mouse clicks, moves and drags, and keyboard presses, releases, and holds.

Answers

Exercise A:

```
glEnable(GL_DEPTH_TEST)

objects = []
cube = Object("Cube")
cube.add_component(Transform((0, 0, -1)))
cube.add_component(Cube(GL_POLYGON, "images/wall.tif"))
cube2 = Object("Cube2")
cube2.add_component(Transform((0, 1, 0)))
cube2.add_component(Cube(GL_POLYGON, "images/brick.tif"))
objects.append(cube)
```

```
objects.append(cube2)
glEnable(GL_DEPTH_TEST)
..
..
while not done:
    ..
    glClear(GL_COLOR_BUFFER_BIT | GL_DEPTH_BUFFER_BIT)
    glRotatef(5, 1, 0, 1)
    for o in objects:
        o.update()
    pygame.display.flip()
```

7

Interactions with the Keyboard and Mouse for Dynamic Graphics Programs

For computer games, graphics applications, and mobile applications, the **mouse** is a key way of interacting with graphics windows. In this case, by mouse, I also include **finger touches** as they are processed in *almost* the same way. The mouse is a pointer that moves across the screen and represents a pixel location (usually at the tip of the arrow if that is the cursor you are using). This location is represented as an (x, y) coordinate in the 2D plane, that is, the graphics window.

Although technically you could create graphics applications without the need for interaction, having these peripherals available becomes useful when testing out an application, moving around in the virtual world, and interacting with models and user interface objects. Hence, I'm adding this chapter early to allow the ability to explore graphics concepts in later chapters with such input. This knowledge will also serve you well as you go forward to make your own custom applications. As such, this chapter is composed of two sections:

- Working with mouse interactions
- Adding keyboard commands

Technical requirements

In this chapter, we will be using Python, PyCharm, Pygame, and PyOpenGL, as used in the previous chapters. Before you begin coding, create a new folder in the PyCharm project for the contents of this chapter called `Chapter 7`.

The solution files containing the code can be found on GitHub at `https://github.com/PacktPublishing/Mathematics-for-Game-Programming-and-Computer-Graphics/tree/main/Chapter07` in the `Chapter07` folder.

Working with mouse interactions

The position of the mouse on the screen is recorded as a pixel location based on the screen coordinate system. As you discovered in earlier chapters, the screen coordinate system by default has its origin in the upper-left corner of the graphics window.

The actions that can be taken with a mouse include the following:

- Pressing down any of the mouse buttons with a finger, including the left, right, and middle buttons
- Releasing a finger press on any of the mouse buttons
- Moving the mouse without pressing any buttons
- Dragging, which is moving the mouse while a button is being held
- Scrolling, in which the user's finger turns the mouse wheel, or on a touchpad or touch mouse, gliding the finger over the surface either vertically or horizontally

The mouse can be used for selecting objects, moving objects, clicking on buttons or input fields, moving the camera, and drawing on the screen. Many of these functions will be explored throughout the book, with the key ones investigated here.

As revealed in *Chapter 1, Hello Graphics Window: You're On Your Way*, one way to capture mouse events is to use code like the following:

```
for event in pygame.event.get():
        elif event.type == MOUSEBUTTONDOWN:
            #do this
```

Pygame can recognize the following mouse events, besides MOUSEBUTTONDOWN:

- MOUSEBUTTONUP
- MOUSEMOTION
- MOUSEWHEEL

The best way to get to know these is to implement them. We used MOUSEBUTTONDOWN in *Chapter 1, Hello Graphics Window: You're On Your Way*, for drawing lines and polygons in the window, so let's look at the others.

Let's do it...

Before you begin, ensure you make a new folder in PyCharm for this code, then follow these steps:

1. Create a new Python script called `MouseDrawing.py` and start with the following code:

```python
import pygame
from pygame.locals import *
pygame.init()
screen_width = 800
screen_height = 800
screen = pygame.display.set_mode((screen_width,
                                  screen_height))
done = False
white = pygame.Color(255, 255, 255)
while not done:
    for event in pygame.event.get():
        if event.type == pygame.QUIT:
            done = True
        elif event.type == MOUSEBUTTONDOWN:
            pygame.draw.rect(screen, white,
                             (pygame.mouse.get_pos(),
                             (5, 5)))
    pygame.display.update()
pygame.quit()
```

When run, this script will allow you to draw a small square at the location of the mouse when you click a mouse button.

2. Modify the code to only draw squares when the left mouse button is pressed with the following:

```python
elif event.type == MOUSEBUTTONDOWN:
    if event.button == 1:
        pygame.draw.rect(screen, white,
                         (pygame.mouse.get_pos(),
                         (5, 5)))
```

Be sure to add an indent after checking for the button value. In this case, you'll be checking for a left-click.

3. You can also check for other buttons using the values from 2 to 5, as follows:

    - 1: Left-click

    - 2: Middle-click

    - 3: Right-click

    - 4: Scroll up

    - 5: Scroll down

    Try it out for yourself.

4. We will now modify the code to draw when the mouse is moved and the left mouse button is pressed down. Change your code like this:

```
mouse_down = False
while not done:
    for event in pygame.event.get():
        if event.type == pygame.QUIT:
            done = True
        elif event.type == MOUSEBUTTONDOWN and \
                event.button == 1:
            mouse_down = True
        elif event.type == MOUSEBUTTONUP and \
                event.button == 1:
            mouse_down = False
        elif event.type == MOUSEMOTION and \
                mouse_down is True:
            pygame.draw.rect(screen, white,
                                (pygame.mouse.get_pos(),
                                (5, 5)))
    pygame.display.update()
```

In the preceding code, a Boolean is created to keep track of the state of the mouse button. Then, the if-else statement is expanded to set the mouse_down value to True when the mouse button is pressed down and then to False when it is released. When you run this, you will be able to draw in the window with the mouse. Notice, though, that if you draw quickly, there will be large gaps between the squares, as shown here:

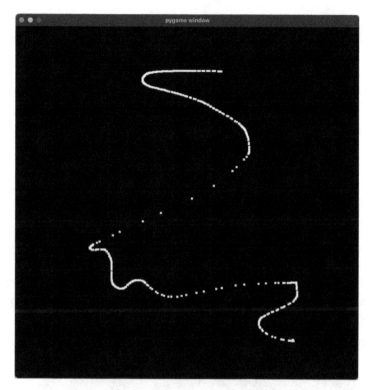

Figure 7.1: Gaps between mouse-drawn pixels

The reason for the large gaps is that you can update the position of the mouse faster than the main loop can keep up with your movements.

> **Note**
>
> Before we continue, it is also interesting to note that the exact logic we discussed in this section is used with finger touches and drags on a mobile device to track the user's finger.

5. To draw a continual line in the window, instead of single squares, we need to draw a line between the last position of the mouse and the next position of the mouse. This will then draw the line to bridge any gaps if the mouse moves too fast for the main loop. This means storing the position of the mouse in the last loop and then drawing a line from the last position of the mouse to the position of the mouse in the current loop. You can do it like this:

```
last_mouse_pos = (0, 0)
while not done:
    for event in pygame.event.get():
        ..
```

```
          elif event.type == MOUSEBUTTONDOWN and \
                      event.button == 1:
     mouse_down = True
     last_mouse_pos = pygame.mouse.get_pos()
 ..
          elif event.type == MOUSEMOTION and \
     mouse_down is True:
     pygame.draw.line(screen, white,
                      last_mouse_pos,
                      pygame.mouse.get_pos(),
                      5)
     last_mouse_pos = pygame.mouse.get_pos()
pygame.display.update()
```

Instead of drawing the square as a point under the mouse, a line is created with the `pygame.draw.line()` function. It takes as parameters the screen, line color, starting pixel position, ending pixel position, and line width.

You will now have the ability to scrawl in white on the window, as shown in *Figure 7.2*:

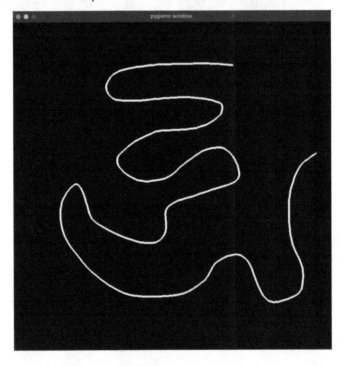

Figure 7.2: No gaps between mouse-drawn pixels

Often, when using a mouse in a graphics environment, the user will want to click on a button or object. This involves calculating whether the mouse position is inside the visual boundaries of the object. For a button, which is basically a rectangle, the mouse position must be inside the range of the button's top-left and bottom-right coordinates, as shown in *Figure 7.3*:

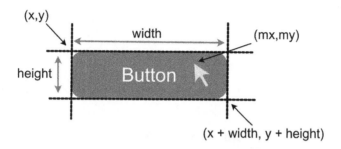

Figure 7.3: The coordinates of a button and a mouse

Given these values, the logic to determine whether the mouse is inside the button boundaries is as follows:

```
if x < mx < (x + width) and y < my < (y + height)
```

Your turn...

Exercise A: Create a green rectangle and add it to the upper-left corner of the window that you have been drawing in. You can do this with the following code:

```
pygame.draw.rect(screen, green, button)
```

where green = pygame.Color(0, 255, 0) and button = (0, 0, 100, 30).

Program the preceding if statement into the existing if-then-else statement inside the main loop to print out **Mouse Over** when the mouse moves over the button.

> **Hint**
>
> To retrieve the mouse position, use mpos = pygame.mouse.get_pos().
>
> You will then be able to set the values of mx and my, as follows:
>
> mx = mpos[0]
>
> my = mpos[1]

Combining 2D and 3D environments

In the previous section, we worked purely with Pygame and a 2D rectangle in 2D space, but what happens if you want to combine a 2D and 3D view? Working with OpenGL means melding an orthographic view for the 2D on top of a perspective view for 3D objects. This can be done by ensuring the correct objects are split to be drawn between projection modes. The best way to explain this is through its application.

Let's do it...

In this exercise, you will learn how to combine a 2D and 3D environment into the same window. Follow these steps:

1. Create a new Python script called `AddingButton.py` and add the following code:

```
from Object import *
from Cube import *
from pygame.locals import *
from OpenGL.GL import *
from OpenGL.GLU import *
pygame.init()
screen_width = 800
screen_height = 600
pygame.display.set_caption('OpenGL in Python')
screen = pygame.display.set_mode((screen_width,
                                  screen_height),
                                 DOUBLEBUF | OPENGL)
done = False
white = pygame.Color(255, 255, 255)
objects_3d = []
objects_2d = []
```

To begin the code, we first create a window in the usual manner. Then, to hold the 2D and 3D objects, two arrays are set up, called `objects_2d` and `objects_3d`:

```
cube = Object("Cube")
cube.add_component(Transform((0, 0, -5)))
cube.add_component(Cube(GL_POLYGON,
                        "images/wall.tif"))
objects_3d.append(cube)
```

```
clock = pygame.time.Clock()
fps = 30
```

Next, a cube object is created and added to the array holding 3D objects before setting up the clock:

```
def set_2d():
    glMatrixMode(GL_PROJECTION)
    glLoadIdentity()  # reset projection matrix
    gluOrtho2D(0, screen.get_width(), 0,
                screen.get_height())
    glMatrixMode(GL_MODELVIEW)
    glLoadIdentity()  # reset modelview matrix
    glViewport(0, 0, screen.get_width(),
                screen.get_height())
def set_3d():
    glMatrixMode(GL_PROJECTION)
    glLoadIdentity()
    gluPerspective(60, (screen_width / screen_height),
                    0.1, 100.0)
    glMatrixMode(GL_MODELVIEW)
    glLoadIdentity()
    glViewport(0, 0, screen.get_width(),
                screen.get_height())
    glEnable(GL_DEPTH_TEST)
```

Then, differing methods are created to handle the different OpenGL setups required to draw 2D and 3D objects:

```
while not done:
    for event in pygame.event.get():
        if event.type == pygame.QUIT:
            done = True
    glPushMatrix()
    glClear(GL_COLOR_BUFFER_BIT | GL_DEPTH_BUFFER_BIT)
    set_3d()
    for o in objects_3d:
        o.update()
    set_2d()
    for o in objects_2d:
```

```
        o.update()
    glPopMatrix()
    pygame.display.flip()
    clock.tick(fps)
pygame.quit()
```

To run this, you will need the Python scripts from the previous chapter for Cube.py, Object.py, Mesh3D.py, and Transform.py.

The lines you need to take note of in the preceding code are the two new methods, set_2d() and set_3d(). Each of these sets its own projection matrix. The set_2d() method needs a call to gluOrtho2D() to create an orthographic projection of the same width and height as the display window, whereas the set_3d() method uses the perspective projection we used before. Basically, the 2D objects will make up the **graphical user interface (GUI)** and operate in a different space from those in the 3D world. We want GUI items to sit on the screen. In both of these methods, you will find glLoadIdentity() used after setting the projection modes. This just reinitializes the views. If you don't do that, every time the projection mode is changed, any settings will compound on what came before.

In addition, in this code, there are two object arrays: one for 3D objects and one for 2D objects. This is to ensure that the correct projections are called before drawing the respective items.

Run this script to find a single cube in the middle of the screen. The translations and rotations that we had in the main script previously have been removed so as to not confuse how things are being drawn, and thus the cube will no longer rotate.

2. It's time to create a button. Make a new Python file called Button.py and add the following code:

```
from OpenGL.GL import *
class Button:
    def __init__(self, screen, position, width,
                    height, color,
        o_color, p_color):
            self.screen = screen
        self.position = position
            self.width = width
            self.height = height
            self.normal_color = color
            self.over_color = o_color
            self.pressed_color = p_color
        def draw(self):
```

```
        glPushMatrix()
        glLoadIdentity()
        glBegin(GL_POLYGON)
        glVertex2f(self.position[0], self.position[1])
        glVertex2f(self.position[0] + self.width,
                self.position[1])
        glVertex2f(self.position[0] + self.width,
                self.position[1] + self.height)
        glVertex2f(self.position[0],
                self.position[1] + self.height)
      glEnd()
    glPopMatrix()
```

In this code, the button class is defined. It takes multiple arguments about the screen, the button position, dimensions, and a variety of colors for the button states. The glVertex2f() method is used with a GL_POLYGON setting to draw a rectangle representing the button.

3. Add the following lines to AddingButtons.py:

```
from Button import *

..

cube.add_component(Cube(GL_POLYGON,
                   "../images/wall.tif"))
                   objects_3d.append(cube)
white = pygame.Color(255, 255, 255)
green = pygame.Color(0, 255, 0)
blue = pygame.Color(0, 0, 255)
button1 = Object("Button")
button1.add_component(Button(screen, (0, 0), 100, 50,
                             white, green, blue))
objects_2d.append(button1)
clock = pygame.time.Clock()
fps = 30

..
```

4. Also, amend Objects.py like this:

```
def update(self, events = None):
    glPushMatrix()
    for c in self.components:
```

```
        if isinstance(c, Transform):
            pos = c.get_position()
            glTranslatef(pos.x, pos.y, pos.z)
        if isinstance(c, Mesh3D):
            c.draw()
        if isinstance(c, Button):
            c.draw(events)
    glPopMatrix()
```

At this point, you can run the AddingButtons script. You will get the cube in the middle of the screen as before and a button in the bottom-left corner, as shown in *Figure 7.4*:

Figure 7.4: A 2D view over the top of a 3D view

You might be wondering two things at this point: *Why isn't the button white?* and *Why is the button at the bottom left of the screen? I thought (0, 0) was at the top left.* Well, the answer lies in the use of the projection view functions.

gluOrtho2d() defined

For the `gluOrtho2d()` method call, the parameters are specified in the following order (as can be found at `https://www.khronos.org/registry/OpenGL-Refpages/gl2.1/xhtml/gluOrtho2D.xml`):

```
void gluOrtho2D(GLdouble left, GLdouble right, GLdouble bottom,
GLdouble top)
```

In the `setup_2d()` method, both the left and bottom of the window are set to 0, meaning we've flipped the *y* axis. Therefore, anything drawn with this setting will assume the origin of the world is at the bottom left of the window. Hence, the button is drawn there.

As for the color of your button, it might not be the same as mine, but it will be a color taken from the texture you have on the cube. Why? Because you added a texture to the cube by turning on OpenGL's texturing mode, but it was never turned off. Therefore, anything drawn afterward will have the same texture, if the texture isn't set to something else.

5. To fix this, go into the `Mesh3D.py` code and add the following:

```
def draw(self):
    glEnable(GL_TEXTURE_2D)
    glTexEnvf(GL_TEXTURE_ENV, GL_TEXTURE_ENV_MODE,
              GL_DECAL)
    glBindTexture(GL_TEXTURE_2D, self.texID)
    for t in range(0, len(self.triangles), 3):
        glBegin(self.draw_type)
        ..
        glEnd()
    glDisable(GL_TEXTURE_2D)
```

Run this. The button will now be white. This might not be the default color you wanted for the button but that's okay as we haven't set it yet.

6. To use the color, specified as the button's normal color when drawing it, modify the button code to the following:

```
def draw(self):
    glPushMatrix()
    glLoadIdentity()
    glColor3f(self.normal_color[0],
              self.normal_color[1],
              self.normal_color[2])
    glBegin(GL_POLYGON)
```

This will draw the button in the color passed through as the normal color. If you are still getting white, then check in `AddingButtons.py` that white isn't the first color you are using in the button creation.

This exercise has demonstrated how to mix 2D and 3D objects in the same scene. As you will have noticed, it's straightforward to switch between different camera projections to achieve the results required when creating 3D applications that have a 2D interface. Of course, now that buttons have been drawn in the window, we will want to add the ability to click on them.

Converting the mouse to projection coordinates

If you put `print(pygame.mouse.get_pos())` into the `while` loop of the application and move the mouse around, you will notice that the projection view settings in OpenGL have had no effect on the mouse coordinates being read by Pygame. Therefore, we need to do a few calculations to determine whether the mouse pointer is over any object we are interested in.

Let's start by looking at the issue in 2D. To ensure we get a full understanding of the task at hand, let's modify the `gluOrtho2D()` call in the `setup_2d()` method to the following:

```
gluOrtho2D(0, 1600 0, 1200)
```

If you make this change and run the application, the button will be half the size it was before as you have now doubled the resolution of the window without changing the actual size of the window. This gives us the situation illustrated in *Figure 7.5*:

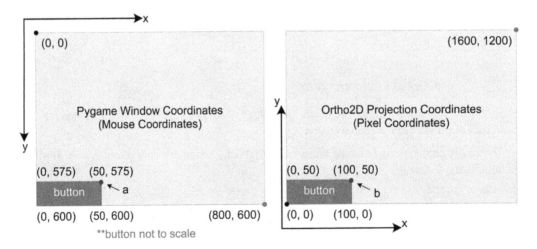

Figure 7.5: The different coordinate systems used in the same window –
one for Pygame coordinates and one for OpenGL Ortho2D

Now, let's see the code that is added to display the button in *Figure 7.5*:

```
button1.add_component(Button(screen, (0, 0), 100, 50,
                      white, green, blue))
```

When the button is now displayed, the coordinates specify for it to start at the origin and be 100 wide by 50 high, position it in the Ortho2D window on the right of *Figure 7.5*. However, these coordinates with respect to the system the mouse lives in are very different, as seen on the left of *Figure 7.5*. This means that before we can determine whether the mouse is over the button as we did earlier, we must either convert the mouse coordinates into Ortho2D space or the button coordinates into the Pygame window space. This is straightforward using a mapping function.

Take points **a** and **b** in *Figure 7.5*, where **a** is in mouse space and **b** is in ortho2d space. The proportions of the x and y values of **a** with respect to the width and height of the mouse window are equal to the proportions of the x and y values of **b** with respect to the width and height of the ortho2d window.

This relationship for the x values can be expressed as follows:

$$\frac{a.x}{mouse\ window\ width} = \frac{b.x}{ortho2d\ window\ width}$$

This relationship for the y values can be expressed as follows:

$$\frac{a.y}{mouse\ window\ height} = \frac{b.y}{ortho2d\ window\ height}$$

Given these relationships, we can devise a mapping function to transform points from one coordinate system into another, as follows:

```
def map_value(current_min, current_max, new_min, new_max,
              value):
    current_range = current_max - current_min
    new_range = new_max - new_min
    return new_min + new_range * ((value-
        current_min)/current_range)
```

Here, `current_min` and `current_max` are the extreme values for the mouse window, for example, the window the mouse coordinates are taken from, and `new_min` and `new_max` are the extremes of the ortho2d window.

Let's look at how you might use this function, which can be used to map values between any two scales. In the case of the current problem, we have a mouse window that has a height extreme of 0 and 600 and an ortho2d window with extremes of 1200 and 0. Note that when specifying the minimum and maximum values, you ensure you are taking the corresponding values. For these two windows, the *y* value for the top of the mouse window is 0, and for the top of the ortho2d window, it is 1200.

To use this in the function to determine the value of b.y, given a.y, it would be written as follows:

```
b.y = map_value(0, 600, 1200, 0, a.y)
```

The same process is required for b.x:

```
b.x = map_value(0, 800, 0, 1600, a.x)
```

Now that we have a method to convert the mouse pointer coordinates into the ortho2d space, we can integrate this into our application to program a mouse-over color change for the button.

Let's do it...

Here, we will create a utility script to hold regularly used and helpful methods:

1. Create a new Python script called Utils.py. We will be using this script to store utility functions as we learn about them. They are useful methods that don't belong to just one class. To this file, add the code for the map_value() function discussed in the previous section.

2. For the Button class to process a mouse over, it needs access to the events picked up in the main script. Therefore, it needs to be passed through the Object class. To do this, begin by making the following changes to AddingButtons.py:

```
while not done:
    events = pygame.event.get()
    for event in events:
        if event.type == pygame.QUIT:
            done = True

    ..
    set_3d()
    for o in objects_3d:
        o.update(events)
    set_2d()
    for o in objects_2d:
        o.update(events)
```

3. Next, Objects.py should be updated as follows:

```
def update(self, events = None):
    glPushMatrix()
```

```
    for c in self.components:
        ..
        if isinstance(c, Button):
            c.draw(events)
    glPopMatrix()
```

4. Last but not least, Button.py is updated with the following:

```
from OpenGL.GL import *
from Utils import *
class Button:

..

def draw(self, events):
    mouse_pos = pygame.mouse.get_pos()
    mx = map_value(0, 800, 0, 1600, mouse_pos[0])
    my = map_value(0, 600, 1200, 0, mouse_pos[1])
    glPushMatrix()
    glLoadIdentity()
    # if mouse over button
    if self.position[0] < mx < (self.position[0] +
        self.width) and \
        self.position[1] < my < (self.position[1] +
            self.height):
        glColor3f(self.over_color[0],
self.over_color[1],
            self.over_color[2]);
    else:
        glColor3f(self.normal_color[0],
                self.normal_color[1],
                self.normal_color[2]);
    glBegin(GL_POLYGON)
```

Most of the preceding code passes through events from the main script in AddingButtons.py. The color change of the button then uses the map_value() function in Utils.py to convert the mouse coordinates in ortho2d space and then sets the color using the same if statement to determine whether the mouse is over a rectangle (in this case, the button).

5. Run the code to see how moving the mouse over the button changes the color of it.

At this point, your project will now successfully mix the 2D view of a simple interface with the 3D environment rendering a cube.

Your turn...

Exercise B: In the drawing application we created earlier in this chapter, we captured a mouse-click event. As it stands, the Button class has a property called pressed_color that defines the color the button should turn if clicked on. It's time to implement that functionality now in this challenge. Update the draw() method of the Button class to display pressed_color when the button is clicked on; otherwise, show normal_color.

Hint of the code

```
from pygame.locals import *
def draw(self, events):
...
if self.position[0] < mx < (self.position[0] + self.width) and \
    self.position[1] < my < (self.position[1] + self.height):
# Enter code here
# Check the left button is pressed then set mouse_down to True.
# If it's released then set to False
# Check mouse_down is True then set pressed_color
# otherwise set over_color
else:
# set normal color
```

If you get stuck, remember that the answers are at the end of this chapter.

Exercise C: There have been some hardcoded values placed in all the code we've been developing, specifically the dimensions of the screen (width and height) and the gluOrtho2d() method. As this isn't best practice, create another Python script file called Settings.py and create variables in there for the dimensions of both window spaces. Then, adjust everywhere in the code that will require the use of these variables. Remember to import Settings.py at the top of each script that uses these variables.

It wouldn't be much of a button if it didn't do anything when it was clicked. In later chapters, we will link up buttons to do more exciting things, but for now, we will just hook the existing one up to a simple function that prints a statement.

Let's do it...

In this exercise, we will add the ability to make a button clickable:

1. In `AddingButtons.py`, add a new function, as follows:

    ```
    objects_3d.append(cube)
    def button_click():
        print("Hello Button")
    white = pygame.Color(255, 255, 255)
    ```

2. Pass this new function to the `add_component()` method when creating a button:

    ```
    blue = pygame.Color(0, 0, 255)
    button1 = Object("Button")
    button1.add_component(Button(screen, (0, 0), 100, 50,
                          green,
                          white, blue, button_click))
    objects_2d.append(button1)
    ```

3. Now, in the `Button` class, accept the passing through of this function pointer with the following:

    ```
    class Button:
        def __init__(self, screen, position, width,
                     height, color, ocolor, pcolor,
                     on_click):
            self.position = position
            ..
            self.on_click = on_click
    ```

4. Then, call this function when the button is clicked on, like this:

    ```
    for e in events:
        if e.type == MOUSEBUTTONDOWN and e.button == 1:
            self.mouse_down = True
            self.on_click()
        elif e.type == MOUSEBUTTONUP and e.button == 1:
            self.mouse_down = False
    ```

This is an easy way to create a pointer to a function. At this point, you can try it out to see **Hello Button** appear in the console each time the button is pressed.

Thus far, we've examined clicking on a 2D button, but what about a 3D object? This is a somewhat more complicated scenario. What is required is a line projected from the camera's near plane into the 3D world and then determining whether the line hits an object. This functionality requires knowledge of vectors and colliders, which we are yet to cover, so it will be left until *Part 2, Essential Trigonometry*.

With the basics of mouse interactions dealt with, it's time to work on keyboard input.

Adding keyboard commands

Once you've worked with mouse events, keyboard events are a breeze. In fact, key presses have fewer events associated with them than a mouse, though you can use multiple keys at a time for complex commands. But basically, a key possesses the down and up events. With respect to graphics environments, keys are used to influence what is going on in the scene. The most common keys in games, for example, are the arrow keys or WASD for moving an object and the spacebar for jump or fire. An example of combination keys would be holding down the *Shift* key while using the arrows to make the object move faster.

In the following exercise, you will program keys to move an object in the 3D environment.

Let's do it...

To move a 3D object, we can use the `Transform` class and modify the position of the model when a key is pressed. To achieve this, follow these steps:

1. In the `Transform` class, we will be adding a new function that will move an object that has that transform as a component. Add the following function:

    ```
    def move_x(self, amount):
        self.position =
            pygame.math.Vector3(self.position.x +
                                amount, self.position.y,
                                self.position.z)
    ```

2. When a key is pressed, we want this `move x()` function to be called. In this case, we will make the left and right arrow keys call it. Add the following lines to `AddingButtons.py`:

    ```
    while not done:
        events = pygame.event.get()
        for event in events:
            if event.type == pygame.QUIT:
                done = True
            elif event.type == pygame.KEYDOWN:
                if event.key == pygame.K_LEFT:
    ```

```
          trans: Transform =
                 cube.get_component(Transform)
          if trans is not None:
              trans.move_x(-0.1)
glPushMatrix()
glClear(GL_COLOR_BUFFER_BIT | GL_DEPTH_BUFFER_BIT)
```

3. The preceding code now requires another function to get hold of the `Transform` component attached to an object. We, therefore, add such a method to the `Object` class like this:

```
def get_component(self, class_type):
    for c in self.components:
        if type(c) is class_type:
            return c
    return None
```

This function will take a class type and then look for it in the component list, returning it to the caller. In this way, the main script can then use the `move_x()` function to add `-0.1` to the object's *x* position, which in this case is the cube.

4. Run this to see how pressing the left arrow key will move the cube to the left.

The movement, rotation, and scaling of an object will, as we progress through the book, be handled by the `Transform` class. This class will also handle the mathematical functions required by each, whether they be instigated by a keyboard command or automated.

Your turn...

Exercise D: Add the ability to move the cube to the right when the right arrow key is pressed. For a list of keycodes, see `https://www.pygame.org/docs/ref/key.html`.

Exercise E: Program the use of the up and down arrow keys to move the cube up and down.

Besides a single press of a key to call a command in the application, keys can also be held down to repeat a command over and over until the key is released. The way we are currently moving the cube is a little cumbersome and often when an object is being moved with the arrow keys, the user prefers to hold the keys down while the object keeps moving until the key is released. Achieving this is very similar to using the mouse to draw a line in that pressing the key down puts the application into a certain state and releasing the key resets that. We can program this in the same way as for holding a mouse button down with a Boolean.

However, a far simpler way is to use the Pygame built-in `get_pressed()` method, as you are about to discover.

Let's do it...

In this task, we will reprogram the key presses to respond when a key is held down using these steps:

1. Modify the AddButtons.py file's while loop, as follows:

```
..
trans: Transform = cube.get_component(Transform)
while not done:
    events = pygame.event.get()
    for event in events:
        if event.type == pygame.QUIT:
            done = True
    keys = pygame.key.get_pressed()
    if keys[pygame.K_LEFT]:
        trans.move_x(-0.1)
    if keys[pygame.K_RIGHT]:
        trans.move_x(0.1)
    if keys[pygame.K_UP]:
        trans.move_y(0.1)
    if keys[pygame.K_DOWN]:
        trans.move_y(-0.1)
    glPushMatrix()
    glClear(GL_COLOR_BUFFER_BIT | GL_DEPTH_BUFFER_BIT)
..
```

The get_pressed() function returns a dictionary of all the keys that are being held down at any one time, allowing for checking multiple key presses. The index in this dictionary is the keycode. In the preceding code, note that the cube's Transform component is set only once (before the while loop). This is more efficient than setting it repeatedly inside the loop. Also, the keys, beginning with the line keys = pygame.key.get_pressed() are checked after the event for loop and not inside it.

2. Run this and see the cube move around the screen when the arrow keys are held down.

In this section, we've covered two methods in the Pygame API that allow for the capture of key events. The first can detect a single key press or release, whereas the second returns all the keys that are pressed during any one main loop. Which one you choose to use will depend on the application.

Summary

In this chapter, we have focused on using the mouse and keyboard to interact with a graphics application. The basic functions available in Pygame have been explored to assist you in getting up and running with these commands. While not all principal components are required to understand mathematics in graphics, knowing these input methods does make the programs we can create to investigate mathematics more engaging. As you progress through the chapters, you will find that the mouse and keyboard will come in handy to demonstrate various key aspects presented in this book. As we move forward, other ways of using these input devices will be explored.

In the next chapter, the mathematics will step up a gear as we start getting into the fundamental concepts underlying most of the mathematics in graphics and games: **vectors**.

Answers

Exercise A:

```
..
green = pygame.Color(0, 255, 0)
mouse_down = False
last_mouse_pos = (0, 0)
button = (0, 0, 100, 30)
while not done:
    pygame.draw.rect(screen, green, button)
    for event in pygame.event.get():
        ..
        elif event.type == MOUSEMOTION:
            mpos = pygame.mouse.get_pos()
            if button[0] < mpos[0] <
                    (button[0] + button[2]) and \
                button[1] < mpos[1] < (button[1] + button[3]):
                print("Mouse Over")
    pygame.display.update()
pygame.quit()
```

Exercise B:

```
class Button:
    def __init__(self, screen, position, width, height,
                 color, ocolor, pcolor):
```

```
        ..
        self.pressed_color = pcolor
        self.mouse_down = False
    ..
def draw(self, events):
    ..    # if mouse over button
    if self.position[0] < mx < (self.position[0] +
                                self.width) and \
        self.position[1] < my < (self.position[1] +
    self.height):
        for e in events:
            if e.type == MOUSEBUTTONDOWN and e.button == 1:
                self.mouse_down = True
            elif e.type == MOUSEBUTTONUP and e.button == 1:
                self.mouse_down = False
        if self.mouse_down:
            glColor3f(self.pressed_color[0],
                      self.pressed_color[1],
                      self.pressed_color[2])
        else:
            glColor3f(self.over_color[0],
                      self.over_color[1],
                      self.over_color[2]);
    else:
        glColor3f(self.normal_color[0],
                  self.normal_color[1],
                  self.normal_color[2]);
    glBegin(GL_POLYGON)
```

Exercise C:

Settings.py

```
# the actual size of the window in pixels
# left, right, top, bottom window_dimensions = (0, 800, 0, 600)
# the resolution of an Ortho2D projection used for drawing a
GUI
```

```
# left, right, top, bottom
gui_dimensions = (0, 1600, 1200, 0)
```

Button.py

```
def draw(self, events):
    mouse_pos = pygame.mouse.get_pos()
    mx = map_value(window_dimensions[0],
                   window_dimensions[1],
                   gui_dimensions[0], gui_dimensions[1],
                   mouse_pos[0])
    my = map_value(window_dimensions[2],
                   window_dimensions[3],
                   gui_dimensions[2], gui_dimensions[3],
                   mouse_pos[1])
```

AddingButtons.py

```
import math
..
pygame.init()
screen_width = math.fabs(window_dimensions[1] -
                         window_dimensions[0])
screen_height = math.fabs(window_dimensions[3] -
                          window_dimensions[2])
pygame.display.set_caption('OpenGL in Python')
..
def set_2d():
    ..
    gluOrtho2D(gui_dimensions[0], gui_dimensions[1],
               gui_dimensions[3], gui_dimensions[2])
    ..
..
```

Exercise D:

```
while not done:
    events = pygame.event.get()
```

```
        for event in events:
            if event.type == pygame.QUIT:
                done = True
            elif event.type == pygame.KEYDOWN:
                if event.key == pygame.K_LEFT:
                    trans: Transform =
                            cube.get_component(Transform)
                    if trans is not None:
                        trans.move_x(-0.1)
                if event.key == pygame.K_RIGHT:
                    trans: Transform =
                            cube.get_component(Transform)
                    if trans is not None:
                        trans.move_x(0.1)
```

Exercise E:

Add these lines to Transform.py:

```
def move_y(self, amount):
    self.position = pygame.math.Vector3(self.position.x,
                                        self.position.y + amount,
                                        self.position.z)
```

Add these lines to AddingButtons.py:

```
if event.key == pygame.K_RIGHT:
    trans: Transform = cube.get_component(Transform)
    if trans is not None:
        trans.move_x(0.1)
if event.key == pygame.K_UP:
    trans: Transform = cube.get_component(Transform)
    if trans is not None:
        trans.move_y(0.1)
if event.key == pygame.K_DOWN:
    trans: Transform = cube.get_component(Transform)
    if trans is not None:
        trans.move_y(-0.1)
```

Part 2 – Essential Trigonometry

This part will take you through the fundamental mathematical skills that underpin the majority of techniques used in graphics and games. As you will discover, graphics applications don't function without the use of triangles and vectors; therefore, a thorough understanding of these constructs is essential for any graphic or games programmer. As you work through this part, the mathematics principles in geometry and trigonometry will be revealed and integrated into the project to set up the environment for not only drawing and lighting 3D objects but also adding the functionality that will animate them.

In this part, we cover the following chapters:

8
Reviewing Our Knowledge of Triangles

If there's one area of mathematics that you want to focus on during your educational journey through graphics and games programming, make sure it is **vector mathematics**. I cannot stress enough how important having a thorough knowledge of this area is to your success in developing your skills in this domain. **Trigonometry**, which examines the relationships between side lengths and angles in triangles, underpins many of the vector functions. As such, we will be exploring the magic of triangles herein.

Initially, for me, vectors and triangles were taught as separate ideas. It wasn't until I started using vectors in my game programming that it suddenly became clear how understanding triangles was such a big part of understanding vectors. Also, vectors are just an extension of the mathematics of triangles.

Triangles would have to be one of the most interesting and useful geometrical shapes. Mathematically, their properties are fundamental to understanding a great proportion of the mathematics in graphics, including vectors, distances, and angles. They are used almost everywhere in defining the geometry of 2D and 3D space and the movement within.

Therefore, to ensure you get the best coverage of all the essential components in this chapter, we will cover the mathematics of similar and right-angled triangles. We will explore the following topics:

- Comparing similar triangles
- Working with right-angled triangles
- Calculating angles with **sine**, **cosine**, and **tangent**
- Investigating triangles by drawing 3D objects

By the end of this chapter, you will have developed the skills to apply your knowledge of triangles to problem-solving in numerous geometric situations, in addition to loading a triangle-based mesh file into your project.

We will begin exploring these topics by looking at the mathematical properties of triangles. To do this, we will examine the properties of similar triangles that underpin the geometry that was used in the projections we explored in *Chapter 4, Graphics and Game Engine Components*.

Technical requirements

The solution files containing the code for this chapter can be found on GitHub at `https://github.com/PacktPublishing/Mathematics-for-Game-Programming-and-Computer-Graphics/tree/main/Chapter08` in the `Chapter08` folder.

For the practical exercise in this chapter, you will require the `teapot.obj` model file, which can be downloaded from `https://github.com/PacktPublishing/Mathematics-for-Game-Programming-and-Computer-Graphics/tree/main/Chapter08/resources`.

Comparing similar triangles

Two triangles are similar when the only difference is the scale, as shown in *Figure 8.1*, where **Triangle (b)** is a scaled-down version of **Triangle (a)**:

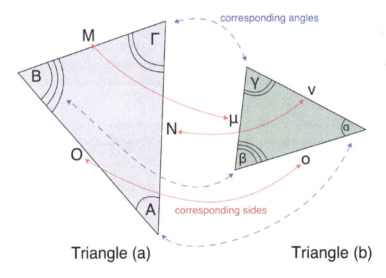

Figure 8.1: Similar triangles

Two triangles are considered similar if they contain the same set of three corner angles. For example, if the angles (in degrees) of **Triangle (a)** are 60, 45, and 75, then the angles of **Triangle (b)** must also be 60, 45, and 75. The total of all angles in a triangle equates to 180 degrees.

For similar triangles, the following applies:

- All corresponding angles are equal

- All corresponding sides have the same ratio

- Corresponding sides have the same ratio between them

As triangles are only considered similar if the respective angles are the same between them, as previously discussed. Given *Figure 8.1*, A = α, B = β, and Γ = γ.

Because, with similar triangles, one is smaller than the other, the ratio of the corresponding sides is the same. For example, in *Figure 8.1*, we have the following:

$$\frac{\mu}{M} = \frac{\nu}{N} = \frac{o}{O}$$

If **Triangle (b)** is half the size of **Triangle (a)**, then we get the following formula:

$$\frac{\mu}{M} = 0.5, \frac{\nu}{N} = 0.5, \text{and } \frac{o}{O} = 0.5 .$$

This is obvious as you would expect that something halved in size would have half the original dimensions.

Let's take a look at an example. Given the triangles in *Figure 8.2*, to calculate the missing length of **Triangle (c)**, first, you need to determine whether both **Triangle (c)** and **Triangle (d)** are similar. In this case, they are similar as they have the same set of angles:

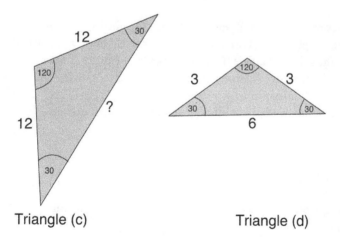

Figure 8.2: Calculating the missing side

Next, you need to determine what the ratio is between the corresponding sides of **Triangle (c)** and **Triangle (d)**. Both given sides of **Triangle (c)** have lengths of **12**. Their corresponding sides in **Triangle (d)**, the ones on either side of the 120-degree angle, are **3**. Therefore, the ratio is 12:3 or 12/3, which equates to 4. We know from similar triangles that the length of the missing side of **Triangle (c)** will give a value of 4 when divided by its corresponding side in **Triangle (d)**, which is **6**:

$4 = ?/6$

Therefore, the length of the missing side is 24.

The same types of comparisons can be used to determine missing angles. Let's take a look at the example in *Figure 8.3*. Here, we can see that there are two triangles with missing angles and sides:

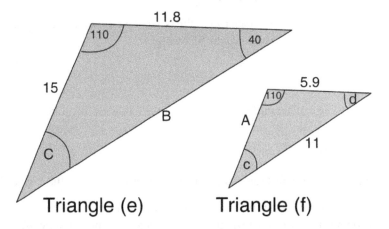

Figure 8.3: Calculating multiple values for similar triangles

How can you tell they are similar? Well, you can't from the information provided, though you could measure the sides and angles to establish that they are similar. But for now, just take my word for it. If you take a look at the side of **Triangle (f)**, which has a length of **5.9**, and its location relative to the angle of **110**, then you'll be able to determine its corresponding side in **Triangle (e)**, which is the side with a length of **11.8**. In this case, the ratio is 11.8/5.9, which equals 2. This tells you that **Triangle (e)** is twice the size of **Triangle (f)**. Given this information, you can calculate the length of **A** like so:

$15/A = 2,$

$A = 7.5$

To determine the length of **B** in **Triangle (e)**, you can use the following calculation:

$B/11 = 2,$

$B = 22$

Finally, there are the missing angles, which are labeled **C**, **D**, and **E**. We can determine angle E as we know that there are 180 degrees in a triangle. So, by subtracting the angles we have in **Triangle (e)**, we can determine the following:

$$C = 180 - 110 - 40 = 30$$

Because the corresponding angles in **Triangle (f)** will have the same value in degrees as **Triangle (e)**, **D** will be 30 and **E** will be 40.

Now, let's put your newfound skills in dealing with similar triangles to use.

Your turn...

Exercise A. Are the triangles in *Figure 8.4* similar? If so, find the value of angle **J** in **Triangle (g)** and the value of angle **K** in **Triangle (h)**:

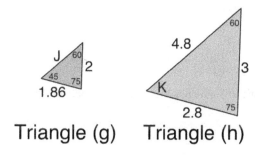

Figure 8.4: Are these similar?

The third observation regarding similar triangles is that the ratios between sides are the same. This means that if you take two sides with the same angle between them in **Triangle (g)** and the corresponding sides in **Triangle (h)**, then their ratios will be equal.

With that, you've learned how similar triangles work and how useful this knowledge was for calculating project spaces in *Chapter 4, Graphics and Game Engine Components*. For these particular calculations, we also took advantage of the properties of a special subset of triangles in which one angle is 90 degrees. These, as you will already know, are called **right-angled triangles**, and understanding them will form the basis for your appreciation of vectors later in this chapter.

Working with right-angled triangles

Since mathematics in grade school, we've learned about **Pythagoras**. He was the Greek mathematician who came up with the relationship between the lengths of the sides of a right-angled triangle. This relationship is so important to vector mathematics that it is worth learning it off by heart. It states that for a right-angled triangle, the square of the **hypotenuse** (the longest side) is equal to the sum of the squares of the other two sides. Such a triangle is illustrated in *Figure 8.5*, with **c** indicating the hypotenuse:

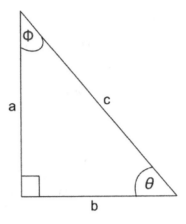

Figure 8.5: A right-angled triangle

Given the right-angled triangle in *Figure 8.5*, the relationship between the hypotenuse and other sides can be written like this:

$$a^2 + b^2 = c^2$$

> **Important note**
> The hypotenuse is the side opposite the right angle.

Let's work through the calculation of the hypotenuse using the triangle in *Figure 8.6*:

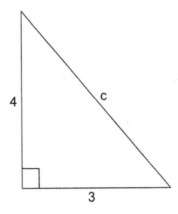

Figure 8.6: A right-angled triangle (with the length of two sides given)

Here, we can see the lengths of two sides of the triangle. With respect to the triangle in *Figure 8.5*, **a** = 4 and **b** = 3. To calculate the length of the hypotenuse (**c**), we must perform the following operations:

$$4^2 + 3^2 = c^2$$

$$16 + 9 = c^2$$

$$\sqrt{25} = c$$

$$c = 5$$

The same formulae can be used to determine the length of a side if the hypotenuse and one side are given (provided you are certain the triangle does indeed have a right angle). Given **Triangle (i)** in *Figure 8.7*, to calculate the side indicated by **x**, we would perform the following:

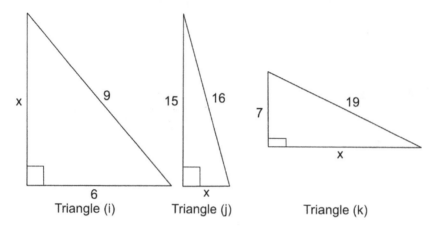

Triangle (i)   Triangle (j)   Triangle (k)

Figure 8.7: Right-angled triangles with a hypotenuse length and one side length

$$x^2 + 6^2 = 9^2$$

$$x^2 + 36 = 81$$

$$x^2 = 81 - 36$$

$$x = \sqrt{45} = 6.7$$

Now, it's your turn to test your knowledge of right-angled triangles.

Your turn...

Exercise B. Calculate the value of **x** for **Triangle (j)** and **Triangle (k)** in *Figure 8.7*.

The hypotenuse of a right-angled triangle, as you will see in *Chapter 9, Practicing Vector Essentials*, represents the body of a vector and it is our new skill in calculating the length of a hypotenuse that will be used over and over again in vector mathematics.

Besides being able to calculate the lengths of the sides of a right-angle triangle by using Pythagoras' theorem, we can also use the value of the angles of the triangle to calculate the same values.

Calculating angles with sine, cosine, and tangent

The length of the hypotenuse and the other two sides can be determined using the triangle's angles and the **sine**, **cosine**, and **tangent** operations. The sides of a right-angled triangle, apart from the hypotenuse, can be labeled as **Opposite** and **Adjacent** based on their positions in the triangle relative to the relevant angle, as shown in *Figure 8.8*:

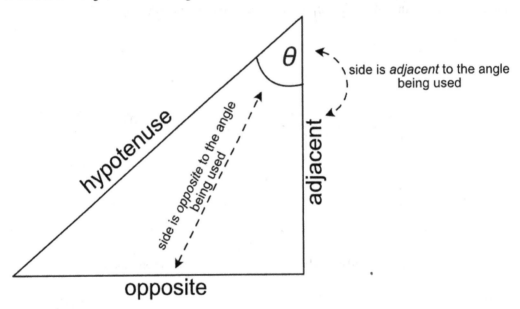

Figure 8.8: The adjacent and opposite triangle sides for θ

Note that for these rules, either angle can be used. If the angle at the other end of the hypotenuse (other than what is shown in *Figure 8.8*) is used, then the labels for **Opposite** and **Adjacent** change positions.

The relationship between the sides and hypotenuse that represent the trigonometric function of sine is as follows:

$$sin\ \theta\ =\ \frac{opposite}{hypotenuse}$$

The relationship for the trigonometric function of cosine is as follows:

$$cos\ \theta\ =\ \frac{adjacent}{hypotenuse}$$

The relationship for the trigonometric function of tangent is as follows:

$$tan \; \theta \; = \; \frac{opposite}{adjacent}$$

Working with the triangles in *Figure 8.5* and *Figure 8.6*, we can calculate angle Φ given that **a** = 4 and **b** = 3. As we have the values for the **Opposite** and **Adjacent** sides, we can use the tangent rule:

$$tan \; \Phi \; = \; \frac{3}{4}$$

$$tan \; \Phi \; = \; 0.75$$

$$\Phi = \; tan^{-1}(0.75)$$

$$\Phi = 0.64 \; radians = 36.87 \text{ degrees}$$

In the preceding calculation, *3* is the length of the side opposite angle Φ and *4* is the length of the adjacent side.

The same calculation can also be calculated with sine:

$$sin \; \Phi \; = \; \frac{3}{5}$$

$$sin \; \Phi \; = \; 0.6$$

$$\Phi = \; sin^{-1}(0.6)$$

$$\Phi = 0.64 \; radians = \; 36.87 \text{ degrees}$$

In this calculation, *3* is the opposite angle and *5* is the hypotenuse. As you may have guessed, we can also perform the same calculation using the cosine method:

$$cos \; \Phi \; = \; \frac{4}{5}$$

$$cos \; \Phi \; = \; 0.8$$

$$\Phi = \; cos^{-1}(0.8)$$

$$\Phi = 0.64 \; radians = \; 36.87 \text{ degrees}$$

Here, because we are now using cosine, the adjacent side of *4* is used instead of the opposite side of *3*.

Note that the trigonometric functions work with radians:

$$1\ rad\ =\ \frac{180^o}{\Pi}\ =\ 57.296^o$$

You can convert between radians and degrees like so:

$$degrees = radians\ \times\ \frac{180^o}{\Pi},$$

$$radians = degrees\ \times\ \frac{\Pi}{180^o}$$

tan-1, *cos-1*, and *sin-1* in the previous equations can be calculated using **arctan**, **arccos**, and **arcsin**, respectively, which are the inverse of each function. If you need to calculate any of these values, using Google is an easy way if you don't have a scientific calculator. For the previous calculation, I found the values by typing `arctan(0.75) in degrees` into my browser's search box. The result is shown in *Figure 8.9*:

Figure 8.9: Using Google to perform calculations

Now, it's your turn to test your newfound skills in determining the lengths of some triangle sides.

Your turn...

Exercise C. Given the right-angled triangles in *Figure 8.10*, find the missing angles and sides:

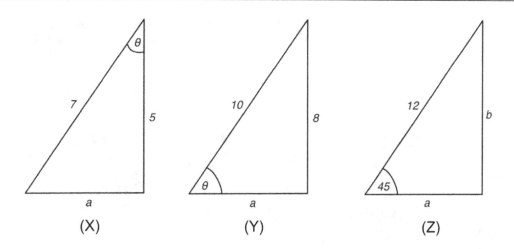

Figure 8.10: Right-angled triangles to test your knowledge of trigonometric functions

Now that you've spent time exploring the mathematics of triangles, it's time to use them in our Python project.

Investigating triangles by drawing 3D objects

Triangles were introduced in our **Python/OpenGL** project in *Chapter 4, Graphics and Game Engine Components*, when they were used to define the vertices and surfaces of 3D models. Triangles are the most efficient way to define a surface as they contain only three vertices. This is the minimum number of points required to define a flat plane with two points only able to define a line, which extends from one to the other.

3D meshes defined with triangles are said to be **triangulated**. While a model can be created by a 3D modeling package with polygons that have more vertices than triangles, software provides artists with a way to triangulate a model so that it can be imported into graphics and game engines. *Figure 8.11* shows a triangulated model of a teapot in **Autodesk Maya LT** and the menu function that will reduce a complex model to triangles:

Figure 8.11: A triangulated teapot in Maya LT

While it is possible to define this teapot model in Python as a manually typed set of vertices, triangles, and UVs, it can be quite laborious to do so. Therefore, before you can explore more complex models in the project we are creating, we should add a simple model loader class.

We will do this by developing a new class that inherits from the Mesh3D class.

Let's do it...

In this exercise, we will examine the format of the Wavefront OBJ 3D model file and develop a Python class to load a model from a .obj file into our project:

1. Create a new folder in PyCharm called Chapter Eight and make copies for it of the final versions of the files from Chapter Seven. Then, include Button.py, Cube.py, Mesh3D.py, Object.py, Settings.py, Transform.py, and Utils.py. Also, make a copy of AddingButtons.py from Chapter Seven for Chapter Eight but rename it DisplayTeapot.py.

2. In the Chapter Eight folder, create a folder called models and copy the teapot.obj file from GitHub into it.

3. Create a new Python script called `LoadMesh.py` and add the following code:

```python
from Mesh3D import *
class LoadMesh(Mesh3D):
    def __init__(self, draw_type, model_filename):
        self.vertices, self.triangles =
                    self.load_drawing(model_filename)
        self.draw_type = draw_type
    def draw(self):
        for t in range(0, len(self.triangles), 3):
            glBegin(self.draw_type)
            glVertex3fv(self.vertices[
                        self.triangles[t]])
            glVertex3fv(self.vertices[
                        self.triangles[t + 1]])
            glVertex3fv(self.vertices[
                        self.triangles[t + 2]])
            glEnd()
        glDisable(GL_TEXTURE_2D)
```

The first part of the `LoadMesh` class declares the vertices and triangles and loads them from a model file that is passed through to the constructor in the form of an external `.obj` file. This will become the `teapot.obj` file in the `models` folder. The `draw()` method is a copy of that from `Mesh3D` with the UV values taken out. They aren't needed initially, so rather than upsetting the existing code in `Mesh3D`, for now, it's simpler to just include another copy here:

```python
    def load_drawing(self, filename):
        vertices = []
        triangles = []
        with open(filename) as fp:
            line = fp.readline()
            while line:
                if line[:2] == "v ":
                    vx, vy, vz = [float(value) for
                                  value in
                                  line[2:].split()]
                    vertices.append((vx, vy, vz))
                if line[:2] == "f ":
                    t1, t2, t3 = [value for value in
```

```
                                      line[2:].split()]
                    triangles.append(
                        [int(value) for value in
                         t1.split('/')][0] - 1)
                    triangles.append(
                        [int(value) for value in
                         t2.split('/')][0] - 1)
                    triangles.append(
                        [int(value) for value in
                         t3.split('/')][0] - 1)
                line = fp.readline()
        return vertices, triangles
```

The next method loads in a .obj file that's looking for the lines starting with 'v ' and 'f '. If you take a look inside the .obj file (you can open it up in PyCharm), you will see that these lines are marked:

```
v 0.000000 2.400000 -1.400000
v 0.229712 2.400000 -1.381970
v 0.227403 2.435440 -1.368070
f 3176/12689/12754 3156/12690/12755 3157/12691/12756
f 3177/12692/12757 3156/12693/12758 3176/12694/12759
f 3177/12695/12760 3155/12696/12761 3156/12697/12762
```

The lines beginning with v are vertex positions. All of these lines collectively define the array of vertices for the mesh.

The lines beginning with f are a set of three indices. The first number in each set specifies the indices in the array of vertex values or what you have been referring to thus far as the triangle values. The other two values are for defining uvs and normals (something we will discuss in future chapters).

A full description of the .obj file format can be found at https://en.wikipedia.org/wiki/Wavefront_.obj_file.

4. To load this model into the project, open DisplayTeapot.py and make the following changes:

```
import math
import pygame.mouse
from Object import *
from Cube import *
from LoadMesh import *
```

```
from pygame.locals import *
...
objects_2d = []
cube = Object("Cube")
cube.add_component(Transform((0, 0, -5)))
cube.add_component(LoadMesh(GL_LINE_LOOP,
                  "models/teapot.obj"))
objects_3d.append(cube)
```

Notice that the code uses the existing cube object and replaces the Cube() class with the LoadMesh() class instead. Now, when you run DisplayTeapot.py, a wireframe teapot will be drawn, as shown in *Figure 8.12*:

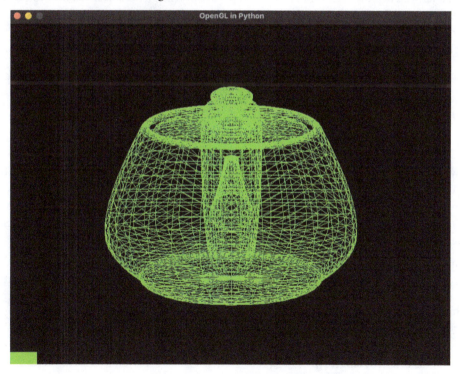

Figure 8.12: A triangulated teapot in Maya LT

The application will run as it did with the cube in *Chapter 7, Interactions with the Keyboard and Mouse for Dynamic Graphics Programs*. You will be able to move the teapot around with the arrow keys.

The OBJ model loader we have added to our project is very basic, but it demonstrates just how many triangles go into making up a mesh. In addition, having a loading functionality in the project will allow you to experiment with bringing other `.obj` files in and loading them. Just ensure they have been triangulated. There's a large variety of other data in the `.obj` file to be dealt with, and we will address this as we progress through this book.

Summary

An understanding of triangles is essential for those who want to program in the graphics and games domains. They are the foundation of the data structures used to store meshes, as well as the basic drawing components of them. When it comes to mathematics in these realms, they are equally as important. Their trigonometric properties, as you will see in the next chapter, form the basis of vector mathematics.

In this chapter, you explored the fundamental properties of triangles and learned how to calculate the length of a triangle's sides and angles using the relationships between similar triangles and the trigonometric rules found in right-angled triangles. Following the theory, we applied our knowledge of triangles to load a triangulated mesh into OpenGL and display it.

Now that you have explored the mathematics of triangles, we can start examining vectors.

Answers

Exercise A:

These two triangles are similar. You only need two corresponding angles and two corresponding sides to determine this fact. Both triangles have angles of 60 and 75. This is enough to establish that the triangles are similar because if you take 60 and 75 away from 180 (the total of the angles in a triangle), then the remaining angle of **K** for **Triangle (h)** will be 45 and the corresponding angle in **Triangle (g)** will also be 45.

To find the length of **J**, you need to establish the ratio of the other corresponding sides. As the triangles are similar, you will know that the known two sides will have the same ratio – that is, 3/2 = 2.8/1.86 = 1.5.

Using this ratio, we can calculate **J** to be 4.8/1.5 = 3.2.

Exercise B:

This is the value of **x** in **Triangle (j)**:

$$x^2 + 15^2 = 16^2$$

$$x^2 + 225 = 256$$

$$x^2 = 256 - 225$$

$$x = \sqrt{31} = 5.57$$

This is the value of **x** in **Triangle (k)**:

$$x^2 + 7^2 = 19^2$$

$$x^2 + 49 = 361$$

$$x^2 = 361 - 49$$

$$x = \sqrt{312} = 17.66$$

Exercise C:

(X) To find θ, we must use the cosine rule, which, when written as a Google search, becomes arccos(5/7) in degrees, which is 44.42. To find the value of **a**, now that we have a value for θ, we can use the sine rule:

$$sin\,(44.42) \;=\; \frac{a}{7}$$

$$sin\,(44.42) \times 7 = a = \;4.9$$

The Google search that we used here is $sin(44.42\ degrees)\ *\ 7$. Note that the value given to the sine function must be in radians, but since we have calculated the angle in degrees, I've simply informed Google that the angle I am giving the sine function is in degrees.

(Y) To find θ, we must use the sine function so that we get arcsin(8/10) in degrees, which gives us 53.13. So, the value of **a** can be found with the cosine rule:

$$cos\,(53.13) \times 10 \;=\; 6$$

(Z) To find **a**, we can use the cosine rule:

$$cos\,(45) \times 12 \;=\; 8.49$$

To find **b**, once we have the value of **a**, we can use the tan rule:

$$tan\,(45) \times 8.49 = 8.49$$

Alternatively, we can use the sine rule:

$$sin\,(45) \times 12 \;=\; 8.49$$

9
Practicing Vector Essentials

If there's one thing you must learn to conquer in the domain of programming graphics (besides triangles), it is vector mathematics. I may have mentioned this before, but that's how important it is! As you will start to see in this chapter and then throughout the rest of this book, you can't do anything in this domain without a solid understanding of them.

This understanding began in *Chapter 8*, *Reviewing Our Knowledge of Triangles*, with the introduction of trigonometry. Vectors are used in everything from defining meshes as the positions of vertices, edges that run between vertices, UV values, and more to moving objects to rendering pixels on the screen. They are extremely versatile, as are their mathematical principles.

This chapter will begin by examining the similarities and differences between the concepts of points and vectors to help you distinguish between the two since, mathematically, they are very similar. We will then go on to discuss the key operations that can be used to manipulate vectors, which will allow you to use them for a variety of graphics programming purposes. Two of these operations that you will see being used over and over again include finding the length of a vector and the angles between two vectors.

To cover these issues, in this chapter, we will cover the following topics:

- Understanding the difference between points and vectors
- Defining vector key operations
- Working out a vector's magnitude
- Exploring the relationship between angles and vectors

The data structure of a vector begins with stored values that have been measured on a Cartesian plane. This plane is a two-dimensional plane formed by the intersection of two axes. Think of it as a large flat surface defined by one of its horizontal edges and one of its vertical edges. While these values are strictly lengths and not precise coordinates, the nature of the data structure, as you will see herein, is most often used in graphics to store points. Though points and vectors cohabit Cartesian space, they are different things. We will discuss this fact in the first section.

By the end of this chapter, you will have a working knowledge of calculating a variety of operations with vectors to position, move, and rotate graphics elements.

Technical requirements

In this chapter, we will be continuing to build on the project that has been developed throughout this book using Python in PyCharm with Pygame and PyOpenGL. The solution files containing the necessary code can be found on GitHub at `https://github.com/PacktPublishing/Mathematics-for-Game-Programming-and-Computer-Graphics/tree/main/Chapter09` in the `Chapter09` folder.

Understanding the difference between points and vectors

We can make sense of vectors by examining them in Cartesian coordinates. A vector in 2D is expressed as (x, y), in 3D as (x, y, z), and in 4D as (x, y, z, w).

Yes, I said four dimensions! At this stage, you are most likely looking at that "w" at the end of the expression and wondering where it came from. Don't worry about it too much as its purpose will become clearer when we examine matrix multiplication.

In theory, a vector can be defined in any number of dimensions extending to infinity. They are used for complex mathematical calculations that can be found in applications relating to machine learning, astrophysics, financial analysis, and inverse kinematics, to name a few. However, in graphics, 2D, 3D, and 4D vectors are used.

Figure 9.1 illustrates a point and a vector in both 2D and 3D space. If you were to just look at the expressions for a vector, shown previously, you could be forgiven for thinking that a vector and point are the same thing. While they are both defined by x, y, z, and other dimensional amounts, this is where the similarities end.

A point is a single location in space. Its coordinates specify its distance along each axis (representative of the dimensions) from the origin or (0, 0, 0) location of the space that the point occupies. For example, the point (3, 5, 3) in *Figure 9.1 (b)* is at a distance of 3 from the origin along the *x* axis, 5 along the *y* axis, and 3 along the *z* axis:

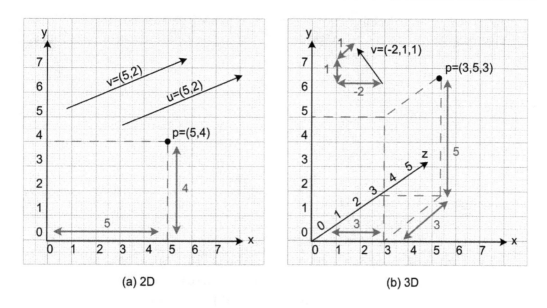

Figure 9.1: A point (p) and a vector (v) shown in (a) 2D and (b) 3D

A vector, however, doesn't specify a location in space. It can be located anywhere. The values that represent it (and look very similar to those of the coordinates of a point) specify the vector's length when broken down into a series of vectors that run parallel with the axis of the space. As shown in *Figure 9.1 (b)*, the v vector is made up of the component vectors that are aligned to the *x*, *y*, and *z* axes. In this case, v is -2 in the x direction, 1 in the y direction, and 1 in the z direction. Even if this vector were moved elsewhere in the space, it would still be represented by (-2, 1, 1). This is illustrated in *Figure 9.1 (a)*, which shows vectors v and u. They are the same.

The thing about vectors is that they don't have a specific starting location. Instead, you must think of them as *instructions*. The best way to understand this is to start working with them.

Let's do it...

In this exercise, we will begin working with vectors by moving the previously drawn cube around the screen. Follow these steps:

1. Create a new folder in PyCharm called `Chapter 9` and make copies of the final versions for it based on the files from `Chapter 8`; include `Button.py`, `Cube.py`, `Mesh3D.py`, `Object.py`, `Settings.py`, `Transform.py`, and `Utils.py`.

2. Make a new Python script called `Vectors.py` and copy into it the final code from the original `AddingButtons.py` file from `Chapter 7`, but *remove* the code that creates the button, as highlighted in the following snippet:

```python
def button_click():
    print("Hello Button")



white = pygame.Color(255, 255, 255)
green = pygame.Color(0, 255, 0)
blue = pygame.Color(0, 0, 255)
button1 = Object("Button")
button1.add_component(Button(screen, (0, 0), 100, 50,
                            green, white, blue,
                            button_click))


objects_2d.append(button1)
```

When this code is run, you will see the regular window and cube in the middle. The cube will still move with the arrow keys.

3. When we start using vectors, we want to be able to measure the results. To do this, we will add a visible grid to the environment. Create a new Python script called `Grid.py` and add the following code to it:

```python
from OpenGL.GL import *

class Grid():
    def __init__(self, interval, halfsize, colour):
        self.interval = interval
        self.halfsize = halfsize
        self.colour = colour

    def draw(self):
        glColor3fv(self.colour)
        glBegin(GL_LINES)
        for x in range(-self.halfsize, self.halfsize):
            for y in range(-self.halfsize,
```

```
                              self.halfsize):
          glVertex3fv((x * self.interval,
                       y * self.interval - 10,
                       0))
          glVertex3fv((x * self.interval,
                       y * self.interval + 500,
                       0))
          glVertex3fv((y * self.interval - 10,
                       x * self.interval, 0))
          glVertex3fv((y * self.interval + 500,
                       x * self.interval, 0))
      glEnd()
```

This new class will be used as another object. When the draw() method is called, the nested for loop will draw a series of horizontal and vertical lines across the screen in the given color. These lines will be spaced by the value of interval.

4. The Grid class can now be used in Vectors.py to draw a grid on the screen:

```
from Grid import *
from Object import *
cube.add_component(Cube(GL_POLYGON,
                   "Chapter_Four/images/wall.tif"))

objects_3d.append(cube)

grid = Object("Grid")
grid.add_component(Transform((0, 0, -5)))
grid.add_component(Grid(0.5, 8, (0, 0, 255)))

objects_3d.append(grid)

clock = pygame.time.Clock()
```

Note that the grid will be transformed back into the screen by -5 since we added a Transform component. Ensure that the cube will move by the same amount.

5. Before running this, we need to tweak `Object.py` to ensure it calls the `draw()` method of the grid. As the call will be the same as for `Mesh3D`, rearrange the `update()` function as follows:

```python
def update(self, events = None):
    glPushMatrix()
    for c in self.components:
        if isinstance(c, Transform):
            pos = c.get_position()
            glTranslatef(pos.x, pos.y, pos.z)
        elif isinstance(c, Mesh3D):
            c.draw()
        elif isinstance(c, Grid):
            c.draw()
        elif isinstance(c, Button):
            c.draw(events)

    glPopMatrix()
```

At this point, run the application; you will see a blue grid across the screen that runs through the center of the cube. The cube will still move with the arrow keys.

6. We will now program the functionality to move the cube using a vector. To the `Transform` class, add a new method called `move`:

```python
def move(self, amount: pygame.math.Vector3):
    self.position = pygame.math.Vector3(
        self.position.x + amount.x,
        self.position.y + amount.y,
        self.position.z + amount.z)
```

This method will accept a 3D vector as the parameter and then use each part of the vector to update the corresponding position coordinate.

7. Let's try using it. In `Vectors.py`, add a key press for the spacebar and move the cube by `0.5` in the x direction, like this:

```python
while not done:
    events = pygame.event.get()
    for event in events:
        if event.type == pygame.QUIT:
            done = True
        if event.type == KEYDOWN:
```

```
if event.key == K_SPACE:
    trans.move(pygame.Vector3(0.5, 0, 0))

keys = pygame.key.get_pressed()
if keys[pygame.K_LEFT]:
```

8. Run the code, take note of where the cube is, and then press the spacebar. The grid lines are `0.5` apart and we are moving the cube by `0.5` in the x direction. The cube will move `0.5` to the right.

    What we have done here is taken the cube's position of (`0, 0, -5`) and added a vector of (`0.5, 0, 0`) to it. This places the cube at a new position of (`0.5, 0, -5`). Each time you press the spacebar, the vector (`0.5, 0, 0`) will be added to the cube's position, resulting in placing the cube at (`1, 0, -5`).

    The position of the cube is a *point*, a location measured from the origin of the world. The movement instructions for the cube of (`0.5, 0, 0`) is a *vector*. This same vector value is added to the position of the cube each time the spacebar is pressed. The vector is not a fixed location in space like the cube's positions. It merely provides instructions for how the cube should move in the x, y and z directions.

As you can see, working with vectors at this level is quite elementary. The x values of the vector affect the x values of a point, the y values of the vector affect the y values of a point, and the z values of a vector affect the z values of a point.

Now, it's your turn to test your understanding of points and vectors.

Your turn...

Exercise A: Modify the code to move the cube by a vector of (`0.5, 0.5, 0.`) each time the spacebar is pressed.

Exercise B: Modify the code to move the cube by a vector of (0, 0, -1) each time the spacebar is pressed.

Having completed this section, you will be starting to understand how easy it is to use vectors to move an object in 3D space. This was achieved using the *addition* operator. We will now take a look at the other vector operations and how they are used in graphics for manipulating points in an environment.

Defining vector key operations

As you discovered previously, the addition of a point (position or location in space) and a vector leads to another point. This was demonstrated by moving the cube from one position to another by adding a vector to the cube's position. We can express this mathematically as:

$P_1 + V = P_2$

Here, *P1* is the initial point, *V* is a vector, and *P2* is the second point. A visualization of this equation is shown in *Figure 9.2*:

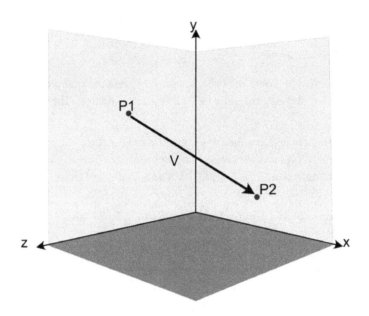

Figure 9.2: The vector, V, when added to point P1 results in point P2

This equation can be rearranged to find the vector between two points:

$$V = P_2 - P_1$$

This equation states that you can find the vector that is required to get from P_1 to P_2 by subtracting P_1 from P_2. Why would you want to do this? Well, if you are given two points and want to know how to get from the first point to the second point, the vector that lies between these points contains the directions that tell you how to get from P_1 to P_2. For example, given the points illustrated in *Figure 9.3*, where *P* is the location of the pirate and *T* is the location of the treasure, the vector, *V*, can be calculated by subtracting *P* from *T*, where the x, y, and z values are subtracted from each other, respectively. These calculations look like this:

$$V = T - P$$
$$= (15, 20, 5) - (9, 6, 2)$$
$$= (15 - 9, 20 - 6, 5 - 2)$$
$$= (6, 14, 3)$$

$V = T - P$

$= (6, 14, 3)$

The vector, V, gives the specifics of how to travel from P to T – that is, 6 in the x direction, 14 in the y direction, and 2 in the z direction. Note that the pirate can travel directly along the vector (as the crow flies) to the treasure or follow each of the constitute x, y, and z components. Either way, we will end up placing the pirate at the location of T. When a vector is split into its components, we can express each as its own vector (one for each axis):

$$V = (6, 0, 0) + (0, 14, 0) + (0, 0, 3)$$

$$= (6 + 0 + 0, 0 + 14 + 0, 0 + 0 + 3)$$

$$= (6, 14, 3)$$

These vectors can be drawn like so:

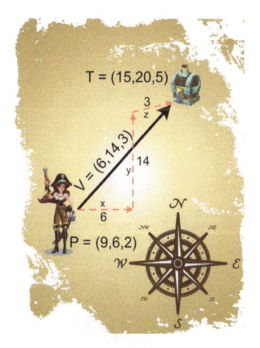

Figure 9.3: The vector between two points

As previously mentioned, a vector has no fixed location in space unless it is applied to a point. The same vector that we've just calculated could be applied to another point. It would then result in a different destination for the pirate. Let's say that if the pirate was located at P = (2, 3, 1), adding the vector V = (6, 14, 3) to that point would place the pirate at (2 + 6, 3 + 14, 1 + 3) = (8, 17, 4) so that they're nowhere near the treasure. If the pirate had moved to P = (2, 3, 1), the vector to the treasure would have to be recalculated as (15, 20, 5) – (2, 3, 1) = (15 – 2, 20 – 3, 5 – 1) = (13, 17, 4). This is a different vector from the previous one.

Vectors can also be added to other vectors to give another vector. This was previously demonstrated in the last equation, where vectors sitting parallel to the x, y, and z axes were added together to give a value for V. But it's not just these vectors that can be added together – any vectors can. The mathematics works the same as for adding vectors and points. For example, given the vectors V = (10, 4, -8) and U = (2, -1, 6), the sum of these is:

$$V + U = (10, 4, -8) + (2, -1, 6)$$
$$= (10 + 2, 4 + -1, -8 + 6)$$
$$= (12, 3, -2)$$

Graphically, this is drawn as shown in *Figure 9.4*. The resulting vector is a vector that extends from the start of the first vector and is drawn as an arrow to the end of the second vector. You can add more than two vectors together. When this is the case, the resulting vector extends from the beginning of the first vector to the end of the last added vector:

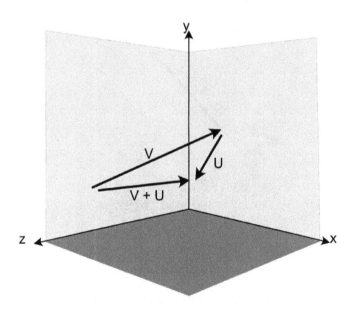

Figure 9.4: Vector addition

Other than addition, there is one other fundamental operation that can be performed on a vector: scaling. When a vector is scaled – that is, multiplied by some value – each of the x, y, and z components is affected. If V = (8, 4, 20), then 2V will be (16, 8, 40) and 0.5V will be (4, 2, 10). We can scale a vector when we need a vector that points in the same direction as the original but isn't as long. If you plot the three preceding vectors, you will see they all point in the same direction. The only difference is their length, as shown in *Figure 9.5*:

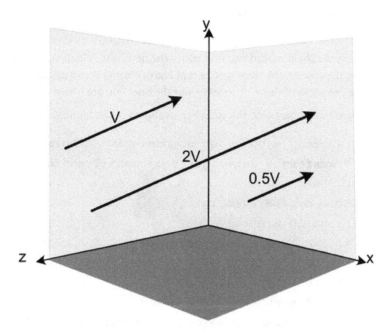

Figure 9.5: Scaled vectors

Working with scaled vectors is no more difficult than what you attempted in the previous exercise, as you are about to discover.

Let's do it...

In this exercise, we will add another cube to the scene we've been working on and use some scaled vectors to investigate how the direction is not affected. Follow these steps:

1. Add another cube to `Vectors.py`, as follows:

```
cube.add_component(Cube(GL_POLYGON,
                    "Chapter_Four/images/wall.tif"))


cube2 = Object("Cube")
cube2.add_component(Transform((0, 1, -5)))
cube2.add_component(Cube(GL_POLYGON,
                    "Chapter_Four/images/brick.tif"))


objects_3d.append(cube)
```

```
objects_3d.append(cube2)
```

Note that the second cube is placed one unit above the first cube. When you run this, you will get two cubes on the screen. The second cube that I have created uses a different texture. Make sure you change the paths of these textures to suit the ones you are using.

2. We will now move both cubes with the spacebar, as shown in the highlighted code:

```
trans: Transform = cube.get_component(Transform)
trans2: Transform = cube2.get_component(Transform)
while not done:
    events = pygame.event.get()
    for event in events:
        if event.type == pygame.QUIT:
            done = True
        if event.type == KEYDOWN:
            if event.key == K_SPACE:
                trans.move(pygame.Vector3(1, 1, 0))
                trans2.move(pygame.Vector3(1, 1, 0) *
                    2)
```

The original cube will move one unit up and one unit to the right. The same vector is being used for the second cube, but it is being multiplied by 2. This will make the second cube move in the same direction but twice as far. Run your application and press the spacebar to witness this.

Notice that the `pygame.Vector3()` class is easy to scale through multiplication.

3. Remember when I said scaled vectors point in the same direction as the original? Well, this wasn't entirely true. What I meant to say was that scaled vectors remain *parallel* to the origin. This is because if we multiply a vector by a negative number, it gets turned around 180 degrees. This is easy enough to see by modifying the move for the second cube, like so:

```
trans2.move(pygame.Vector3(1, 1, 0) * -2)
```

Try this out to see the second cube move in the opposite direction.

In this exercise, we looked at how easy it is to scale a vector. Exactly why this is useful will be made more apparent when we start moving objects around on the screen through animation later in this book.

As vectors are used to move objects or define the distance between one location and another, it is also possible to measure their length, which is another important mathematical function to learn.

In the next section, you will discover how to use Pythagoras' theorem to calculate it.

Working out a vector's magnitude

By drawing and working with vectors, it's obvious to see they have a length. This is something that also distinguishes them from a point. The magnitude or length of a vector is useful for calculating the distance between where it starts and where it ends. For example, in *Figure 9.3*, we calculated the vector the pirate had to travel to get from a starting location to the treasure. From this vector, we can see the direction of travel, but we can also calculate how far the pirate is from the treasure. A very common operation to perform in graphics when it comes to moving objects is determining how far objects are apart, as well as working with collisions and a multitude of other functions. Therefore, it's useful to understand how the magnitude is calculated.

To perform this operation, we must go back to Pythagoras theorem and triangles. Essentially, every vector can be made into a right-angled triangle. From there, the vector itself represents the hypotenuse of the triangle, and as such, the length can be determined. Let's begin by examining an example in 2D. As shown in *Figure 9.6*, any vector can be turned into a right-angled triangle. The lengths of the sides of the triangle relate to the vector's constitute x and y dimensions. For example, the vector (3, 4), when used to construct a right angle, will have sides that are 3 and 4 in length. The magnitude can then be found using Pythagoras theorem:

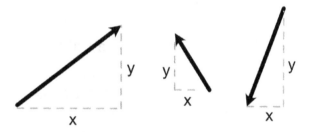

Figure 9.6: Vectors made into right-angled triangles

To calculate the magnitude, use the following formula:

$$x^2 + y^2 = magnitude^2 \text{ or}$$

$$|v| = \sqrt{x^2 + y^2}$$

Here, $|v|$ is the way to mathematically write *the length of v*.

This can be extended to vectors in 3D by adding in a z component:

$$|v| = \sqrt{x^2 + y^2 + z^2}$$

For example, the magnitude of the vector (9, 2, 5) will be:

$$|v| = \sqrt{9^2 + 2^2 + 5^2}$$

$$|v| = \sqrt{81 + 4 + 25}$$

$$|v| = \sqrt{110}$$

$$|v| = 10.5$$

Your turn...

Exercise C: What is the length of the vector (2, 3, 4)?

Having found the connection between triangles and vectors, we are now able to make a connection between vectors and angles. It's the fact that vectors can be expressed as a right-angled triangle that makes this possible. This means we can bring the trigonometric functions of sine, cosine, and tangent, as explained in *Chapter 8*, *Reviewing Our Knowledge of Triangles*, into play.

So far, you've learned how to work with vectors to move from one location to another, as well as how to find the distance between those locations.

In the next section, we will look at another fundamental operation of vectors that will become very useful for rotating graphics elements, as well as providing simple artificial intelligence with turning functionality.

Exploring the relationship between angles and vectors

The angles that are made from intersecting two vectors can best be explored by considering how a game character facing one vector should turn to be facing another. In this section, we will explore how such a game character might turn toward the desired object.

At the end of *Chapter 8*, *Reviewing Our Knowledge of Triangles*, you added a mesh loading class to your project and used it to display a teapot in the graphics window. How much did you want to turn this teapot around to get a better look at it? Besides translation, rotation is another primary transformation used in graphics. Rotations require angles; these angles are measured between vectors using the principles we explored in *Chapter 8*.

For example, let's consider the case shown in *Figure 9.7*, in which the game character, *Chomper*, would like to turn and face the jar of *Vegemite*. A vector is used to represent the direction Chomper is facing, as well as the direction he would like to face, that being the vector from Chomper to the Vegemite:

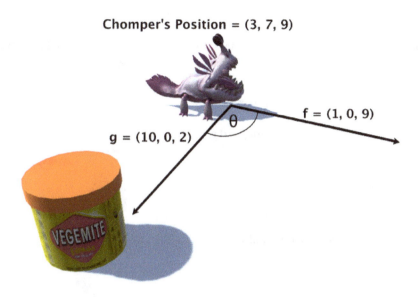

Chomper's Position = (3, 7, 9)

f = (1, 0, 9)

θ

g = (10, 0, 2)

Figure 9.7: Turning a game character to face an object

From the information given, we can conclude that the Vegemite is located at the following position:

Vpos = Chomper_Pos + g

This is because adding a vector to a position gives another position, and we know where Chomper is located and the vector to the Vegemite, given in *Figure 9.7*. Therefore, the Vegemite is located at the following position:

Vpos = (3, 7, 9) + (10, 0, 2)

Vpos = (13, 7, 11)

This is inconsequential to finding the value of θ, which is the angle Chomper has to turn at to face the Vegemite. However, these calculations do demonstrate where positions or points in space and vectors part ways. It doesn't matter where Chomper is located in space – if the Vegemite is relatively in the same position, then the vector to the Vegemite will be the same. Essentially, you can move Chomper and the Vegemite anywhere in space and so long as they are moved by the same amount, the vectors of g and f will remain the same.

So, the question remains, what is the angle Chomper needs to turn at to be facing the Vegemite? To calculate this, we have another interesting operation. Enter the dot product operation!

The dot product

The **dot product** is a mathematical operation that's performed on two vectors. It takes each element of two vectors, multiplies them together, and then adds them together. For the vectors v and w, the dot product is expressed like this:

$$v \cdot w = v.x \times w.x + v.y \times w.y + v.z \times w.z$$

Therefore, the dot product for *Figure 9.7*, for f and g, is like so:

$$f \cdot g = 1 \times 10 + 0 \times 0 + 9 \times 2$$

$$f \cdot g = 10 + 0 + 18$$

$$f \cdot g = 28$$

No, this isn't the angle between vectors f and g, but it's a start. To calculate the angle, θ, between two vectors, we require the following equation:

$$\cos \theta = \frac{f \cdot g}{|f| \times |g|}$$

If you are interested in the derivation of this formula, you are encouraged to read `https://www.mathsisfun.com/algebra/vectors-dot-product.html`.

Given we've already calculated the dot product, we only need to find the magnitude of f and g to use in the divisor (can you work this out from what we covered in the previous section?). The equations with these values will now look like this:

$$\cos \theta = \frac{28}{9 \times 10.2}$$

$$\cos \theta = 0.3$$

$$\theta = cos^{-1}0.3$$

$$\theta = 1.3 \; radians = 72.54 \; degrees$$

From this, we know that Chomper must turn 72.54 degrees from the direction it is facing to then be facing the Vegemite. This is useful information when programming an artificially-intelligent game character. The only issue that remains is to determine whether Chomper should turn clockwise or anticlockwise. For you and me, we can see Chomper should turn clockwise. However, for a game character, this needs to be calculated. Hence, we require a cross-product calculation.

The cross product

The **cross product** of two vectors produces another vector that sits at 90 degrees, otherwise called **perpendicular**, to the original two vectors. For example, if you consider the X, Y, and Z axes of a 3D Cartesian space, then the cross product of the x axis and the y axis would be the z axis. This may not immediately seem like a solution to determining which way Chomper should turn (a problem from the previous section), but it will soon.

When it comes to working on a cross product between two vectors, v and w, the equation is written like this:

$$v \times w$$

This is not a simple multiplication as vectors contain multiple parts. The actual equation for vectors in 3D is this:

$$v \times w = (v.y * w.z - v.z * w.y) + (v.z * w.x - v.x * w.z) + (v.x * w.y - v.y * w.x)$$

Again, if you are interested in the derivation, you can find it at `https://www.mathsisfun.com/algebra/vectors-cross-product.html`. Note that in the preceding equation, I used * to represent a normal multiplication as opposed to the x used in the cross-product equation. I've no doubt the cross-product equation looks pretty scary to most beginners at this point. However, it's an important calculation and if you need to Google it now and again, then so be it.

What is more important is that you understand why you are using it and can evaluate the results it gives you to work with. The result will be another vector. If you plot or visualize the three vectors involved, the result will sit perpendicular to the other two.

Taking the example we've been working through with Chomper and the Vegemite, the cross product of f and g from *Figure 9.7* will be:

$$f \times g = (f.y * g.z - f.z * g.y) + (f.z * g.x - f.x * g.z) + (f.x * g.y - f.y * g.x)$$
$$f \times g = (0 * 2 - 9 * 0) + (9 * 10 - 1 * 2) + (1 * 0 - 0 * 10)$$
$$f \times g = (0) + (88) + (0)$$

Something that I left out in the previous equation and description is that you must end up with three values added together. These represent an x, y, and z component. The complete equation for cases in 3D is:

$$v \times w = (v.y * w.z - v.z * w.y)(1, 0, 0) +$$
$$(v.z * w.x - v.x * w.z)(0, 1, 0) +$$
$$(v.x * w.y - v.y * w.x)(0, 0, 1)$$

This assigns each bracketed equation to x, y, and z, respectively.

The result of all these calculations gives us:

$$f \times g = (0, 88, 0)$$

This is another vector. When drawn into the illustration of *Figure 9.7*, it will give the vector shown in *Figure 9.8*:

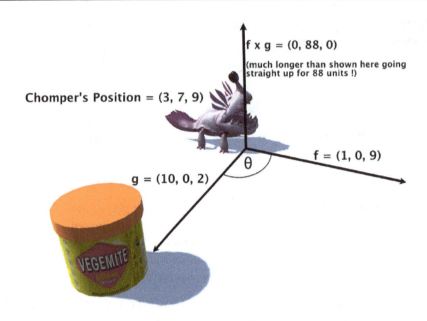

Figure 9.8: The cross product

What does this tell us about the direction that Chomper needs to turn? Well, there are two cross products you can calculate – *f x g* and *g x f* – and they give differing results. Try the calculations for yourself. *f x g* will equate to (0, 88, 0) and *g x f* will equate to (0, -88, 0). As you can see, a different ordering of the vectors in the cross product will produce vectors that are parallel but going in different directions. And this is what we can take advantage of when deciding the direction in which Chomper should turn. In *Figure 9.8*, Chomper should turn clockwise. From that, we can conclude that if the *y* value of the cross product is positive, then Chomper turns clockwise; otherwise, Chomper turns anticlockwise.

In this particular case, Chomper is turning around its Y, or vertical, axis. Knowing the axis of rotation for these calculations is important. For example, if the character needed to turn around its Z-axis, then it would be the negative and positive results from the cross products that would indicate the direction of turn.

All of these properties regarding vectors will be explored in *Chapter 15, Navigating the View Space*, when we start to produce game objects that can self-navigate in the 3D space we've been creating.

Summary

There's a lot to learn when it comes to vector mathematics, and this chapter has only scratched at the surface and plucked out the essential parts relevant to graphics and games. If you get a chance, I would strongly suggest that you review the extra links given throughout. Vector mathematics is a university-level topic. I've attempted to reveal the simplicity of the topic and reduce it to simple addition and multiplication operations, but if you feel you haven't grasped the topic, I'd encourage you to investigate what's available on *Khan Academy* (khanacademy.org). In this book, I'm addressing the most relevant aspects of all mathematical topics concerning computer graphics and games, but unfortunately, due to page count, I can't cover everything.

We started this chapter by looking at the differences between points and vectors, followed by several key vector operations required to manipulate them to move objects in space. This movement was further defined with a distance of movement, which we discovered could be calculated using Pythagoras' theorem. Finally, by introducing the dot and cross-product operations, you learned how to turn a character from facing in one direction to facing toward a goal object. With this new set of skills, you are now well equipped to create animated graphics environments and simple artificially-intelligent characters.

To help forge a place for vectors in your game development arsenal, in the next chapter, we will focus on the aspects of the domain that rely on these fundamental concepts and further assist you in developing your skills and focusing on them in this area. To begin, you will discover the variety of uses of lines, line segments, rays, and normals, and discover how each of these can be defined by the use of vectors before we continue to develop an understanding of how each of these concepts is critical knowledge in graphics programming. You will apply them by animating the movement of cubes and defining the exterior sides of polygons to optimize rendering.

Answers

Exercise A:

```
trans.move(pygame.Vector3(0.5, 0.5, 0))
```

The cube will move diagonally to the top right.

Exercise B:

```
trans.move(pygame.Vector3(0, -1, 0))
```

The cube will move into the screen, away from the viewer.

Exercise C:

$$\bar{v} = \sqrt{2^2 + 3^2 + 4^2}$$

$$\bar{v} = \sqrt{4 + 9 + 16}$$

$$\bar{v} = \sqrt{29}$$

$$\bar{v} = 5.4$$

10
Getting Acquainted with Lines, Rays, and Normals

Since we covered vectors in the previous chapter, it's time to jump into exploring the mathematics of **lines**, **rays**, and **normals**. All three can be defined by vectors and on the surface, all appear to be the same and yet they have very specific and differing uses in graphics. Fortunately, their similarities concerning geometric structure allow them to be defined and manipulated by the same equations, as you will discover in this chapter.

The one thing lines, rays, and normals have in common is that they are all straight. This makes them very useful in graphics for defining space, direction, the edges of meshes, distance, collisions, reflections, and much more. The line construct is one of the fundamental drawing operations in graphics, as we covered in *Chapter 1*, *Hello Graphics Window: You're On Your Way*, and *Chapter 2*, *Let's Start Drawing*.

In this chapter, we will cover the essential knowledge you need to define and manipulate lines, rays, and normals mathematically and build data structures for each into our project. We will begin with a thorough examination of the basic concepts and explore the differing uses for lines, line segments, rays, and normals. Following this, the parametric form of lines will be presented, which you will use to animate the movement of a cube from one location to another in 3D space. We will conclude this chapter by using our practical skills of drawing lines to display the normals of polygon surfaces, which is critical knowledge in understanding how surfaces are textured and lit.

To this end, in this chapter, we will cover the following topics:

- Defining lines, **segments**, and **rays**
- Using the **parametric** form of lines
- Calculating and displaying **normals**

By the end of this chapter, you will have gained a firm understanding of how these elements are used throughout graphics and be able to apply them to your projects.

Technical requirements

In this chapter, we will continue working on the project that we've constructed using Python, PyCharm, Pygame, and PyOpenGL.

The solution files containing this chapter's code can be found on GitHub at `https://github. com/PacktPublishing/Mathematics-for-Game-Programming-and-Computer-Graphics/tree/main/Chapter10` in the `Chapter10` folder.

Defining lines, segments, and rays

What most people call a line is a **line segment**. A line segment is just a piece of a line. By true mathematical definition, a line continues infinitely. In *Chapter 2, Let's Start Drawing*, we used the equation for a line to draw segments between mouse clicks in our project window. Recall that the following equation was used:

$y = mx + c$

Here, m is the **gradient** (or **slope**) and c is the **y-intercept**, as shown in *Figure 10.1*:

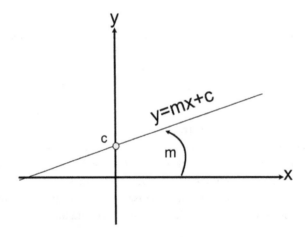

Figure 10.1: The gradient and y-intercept

The value of y can be calculated for infinite values of x and vice versa, making a line continuous. However, a line segment has a start and end, as illustrated in *Figure 10.2*:

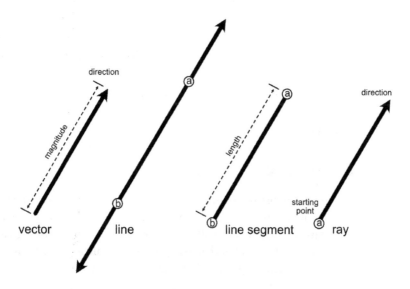

Figure 10.2: Differing straight geometrics

To clarify the difference between these straight geometrics, let's examine their properties:

- A vector has a magnitude and direction, as we discovered in *Chapter 9, Practicing Vector Essentials*. It doesn't have a location in space. Vectors are used in graphics for all manner of operations, including moving objects, measuring distances, and calculating lighting.

- A line continues to infinity from both ends. It can be defined as going through two points. A line occupies a location in space. In reality, lines aren't used in graphics at all as it's not possible to calculate values to infinity. However, lines can be used to calculate segments anywhere along their length for use in drawing and other calculations.

- A line segment can be defined by the same basic equation as a line but is a finite length cut off at both ends. When writing its equation, these cut-off points are featured:

$$y = mx + c \quad b.x \leq x \leq a.x$$

  This formula specifies that x is restricted to values between the x values of the cut-off points – that is, a and b. Although you might think that lines are used everywhere in graphics, line segments are more common as we don't need to calculate straight paths to infinity. A line segment is what you see drawn on the screen and is the basic element used to draw the edges of polygons and meshes.

- A ray has a direction, like a vector, but it continues infinitely in one direction and has a definite starting point. Rays are used to specify the position and direction of lights and calculate collision detection. Rays can be defined by a starting point and using the direction of a vector.

Conceptually, you might think of lines and line segments collectively by the term *line*. It probably doesn't make too much difference if that is how you understand them, so long as you grasp the concepts and can see how each is applied in graphics. I've already been guilty of calling them all lines, so I will continue to do so.

While lines, line segments, rays, and normals might seem fundamentally the same, you've now discovered how they are different. In many graphics calculations, it's rare to use any of these elements alone, as you are about to find out.

Using the parametric form of lines

While the line equation given in the previous section is something most people are familiar with from high school mathematics, it's not particularly useful in graphics when you want to manipulate objects or work out intersections, animations, and collisions. Therefore, we tend to use the parametric form. The parametric form of an equation, rather than using *x* and *y* to calculate positions, uses *time*, represented by *t*. This might sound confusing at first but bear with me while I explain.

Consider *Figure 10.3 (a)*. Notice how a line segment can be represented by two points and a vector going between them:

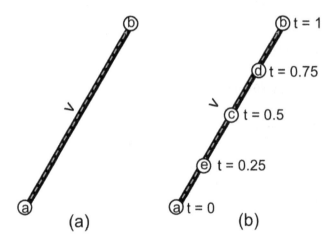

Figure 10.3: A line segment with a vector between the start and end points

The calculation for *v* is as follows:

$$v = b - a$$

We can also express it like so:

$b = a + v$

This tells us that if we start at point *a* and travel along the whole length of *v*, we will end up at *b*. Where would you be if you only traveled halfway along *v*? This can be expressed as $b = a + 0.5v$. Here, you'd be at point *c*, as shown in *Figure 10.2 (b)*. If you traveled no distance of *v*, then you'd still be at *a*.

The parametric form of an equation uses this scalar of *v* as the variable, *t*. So, at *a*, $t = 0$ and at *b*, $t = 1$. Therefore, the parametric equation for a line segment is as follows:

$b = a + vt \quad 0 \le t \le 1$

Now, just because *t* is restricted to values of 0 and 1 here doesn't mean it has to be this way. If you wanted to define a line using the parametric form, you could express it like this:

$b = a + vt \quad \infty \le t \le \infty$

This means that *t* can be any value at all. However, when it is 0, you'll still have *a*, and when it is 1, you'll still have *b*. However, plugging in any other value for *t* will provide points along the line that extend to infinity. Therefore, *t* represents progress along the line from *a* to *b*, which specifies the time traveled along the vector.

Your turn...

- *Exercise A:*

  - What are the coordinates of the end point of a line segment where the start location is (3, 6, 7) and the vector from the start location to the end point is (1, 5, 10)?

- *Exercise B:*

  - What are the coordinates halfway along the line segment going from (1, 9, 5) and (10, 18, 3)?

- *Exercise C:*

  - What are the coordinates of a point 75% along a line segment where the end position is (2, 3, 4) and the vector from the start to the end is (5, 5, 3)?

Now, let's use a parametric equation to linearly interpolate the position of a cube in our project to animate movement.

Let's do it...

In this exercise, we will animate a single cube so that it moves along a line between two points using a parametric equation. This will further reinforce the use of the variable, *t*, to represent time:

1. Make a new folder in PyCharm called `Chapter 10`. Make a copy of all the files from the `Chapter 9` folder and place them in this new folder. This includes `Vectors.py`.

2. Rename the copy of `Vectors.py` to `Animate.py`.

3. Modify the content of `Animate.py`, like this:

```
. .
        glViewport(0, 0, screen.get_width(),
                          screen.get_height())
        glEnable(GL_DEPTH_TEST)


    trans: Transform = cube.get_component(Transform)
    start_position = pygame.Vector3(-3, 0, -5)
    end_position = pygame.Vector3(3, 0, -5)
    v = end_position - start_position
    t = 0
    trans.set_position(start_position)
    dt = 0

    trans2: Transform = cube2.get_component(Transform)
    while not done:
        events = pygame.event.get()
        for event in events:
            if event.type == pygame.QUIT:
                done = True

        if t <= 1:
            trans.set_position(start_position + t * v)
            t += 0.0001 * dt

        glPushMatrix()
        glClear(GL_COLOR_BUFFER_BIT | GL_DEPTH_BUFFER_BIT)
. .
```

```
        glPopMatrix()
        pygame.display.flip()
        dt = clock.tick(fps)
    pygame.quit()
```

Here, we initially set the position of the first cube to the left-hand side of the screen, as shown in *Figure 10.4*, as well as after setting up the parametric equation values, including calculating v and initializing t. For this, we defined a start and end position for the movement.

There's also a new value in there called dt. This will record the time taken for a single game loop. It gets its value from the `clock.tick()` function at the end of the game loop. Multiplying the movement by it (in this case, the updating value of t) will ensure the speed of movement is consistent across differing computers.

The `set_position()` function of the `Transform` class is used to set the changing position of the cube by feeding in the parametric equation.

Take note that the code for moving the cubes with the arrow keys and spacebar has been removed:

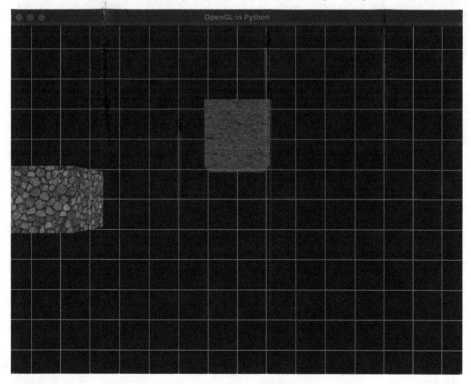

Figure 10.4: Starting position of the animated cube

4. Press play and watch the cube move across the screen, stopping when t = 1, as shown in *Figure 10.5*:

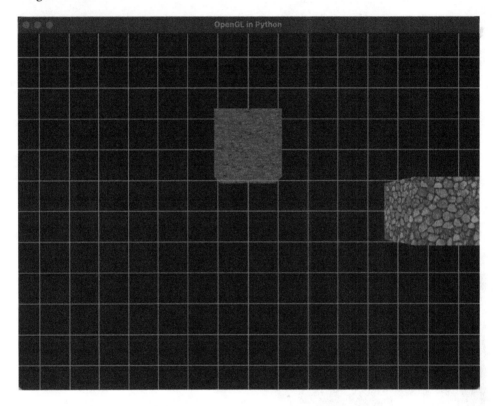

Figure 10.5: Final position of the animated cube

5. Now, modify the ending position of the line segment to (0, 0, -5), like this:

```
end_position = pygame.Vector3(0, 0, -5)
```

6. Run the program again. The cube will stop in the center of the screen.

7. To see how values of t greater than 1 will keep the cube moving along the same line, comment out the if statement, as follows:

```
#if t <= 1:
```

8. Run the program one more time. The cube will move across the screen and out of view on the right-hand side.

    The code you've created in this exercise will move the cube between any two points, but make sure you put the if statement we took out previously back in.

As you've experienced in this section, whether we are working with points, lines, or vectors, the mathematics is the same. Sometimes, it may get confusing as to what is a vector and what is a point; the same goes for lines and line segments. Therefore, a thorough understanding of context is essential; otherwise, it just appears as though you are adding and subtracting *x, y,* and *z* values.

Other special vector-related *lines* (I mean this in the general sense) are normals and rays. As you will see, they too have a special place in graphics.

Calculating and displaying normals

Unlike rays, which have a starting position and direction and are infinite in length, normals are special vectors assigned to mesh surfaces and vertices. A normal is a vector that usually sits at 90 degrees to a surface. I say *usually* because they can be manipulated for special surface texturing effects. Normals are used to calculate how light falls on a surface, as well as define the side of a polygon to which a texture is applied. There are two places normals are used; on surfaces and vertices, as shown in *Figure 10.6*, though usually, they are defined with vertices in mesh files:

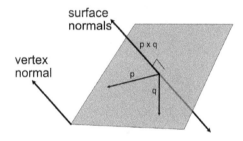

Figure 10.6: Normals

As shown in *Figure 10.6*, mathematically, a plane has two normals, which can be found using the cross product of two vectors on the surface. Recall from *Chapter 9, Practicing Vector Essentials,* that multiplying two vectors together results in a third vector that sits at right angles to the initial vectors. In *Figure 10.6*, one of the surface normals can be found with *q* x *p* and the other with *p* x *q*. In graphics, there's usually only a normal on one side of a plane. This helps define the front side of a polygon. *Figure 10.7* shows a mesh in Maya displaying the surface and vertex normals:

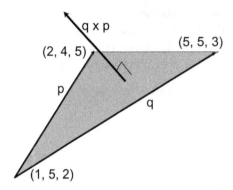

Figure 10.7: Face and vertex normals on a mesh

To calculate a normal, given a triangle that represents one of the flat surfaces in a mesh, two vectors that run between the vertices can be calculated to find vertexes that sit flush with the surface. Then, they can be crossed together to get the normal.

For example, given the triangle in *Figure 10.8*, the vectors of *p* and *q* can be calculated using the vertex values:

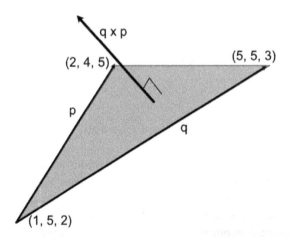

Figure 10.8: Calculating the normal of a triangle

Here, we have the following:

p = (2, 4, 5) – (1, 5, 2) = (1, -1, 3)

q = (5, 5, 3) – (1, 5, 2) = (4, 0, 1)

q x p = (q.y * p.z – q.z * p.y), (q.z * p.x – q.x * p.z), (q.x * p.y – q.y * p.x)

q x p = (1, -11, -4)

Note that the normal is a vector, so it does not need to originate anywhere special on the surface. It only has a direction and magnitude.

Your turn...

- *Exercise D*:

  - What are the normal vectors for the vectors v = (8, 2, 3) and w = (1, 2, 3)? What observation can you make about the values contained in the two normals?

- *Exercise E*:

  - What are the normal vectors for the triangle defined by the vertices a = (3, 1, 2), b = (9, 5, 2) and c = (15, 8, 7)?

Let's try calculating and displaying some normals for the cubes we've been working with throughout this project.

Let's do it...

In this exercise, we will take some of the vertices of the cubes we've been working on and calculate the normals to display. Follow these steps:

1. Create a new Python script called MathOGL.py. We will use this file to code some mathematical functions.

2. Let's start by adding functions to MathOGL.py with a cross-product calculation, like this:

```python
import pygame


def cross_product(v, w):
    return pygame.Vector3((v.y*w.z - v.z*w.y),
                          (v.x*w.z - v.z*w.x),
                          (v.x*w.y - v.y*w.x))
```

You'll recognize the cross-product equation we examined in *Chapter 9, Practicing Vector Essentials*.

3. Create a new Python script called `DisplayNormals.py`. This file will house a new class that can be added to an object to display the normals.

4. Add the following code to `DisplayNormals.py`:

```python
import pygame
from OpenGL.GL import *
from MathOGL import *


class DisplayNormals:

    def __init__(self, vertices, triangles):
        self.vertices = vertices
        self.triangles = triangles
        self.normals = []
        for t in range(0, len(self.triangles), 3):
            vertex1 = self.vertices[self.triangles[t]]
            vertex2 = self.vertices[self.triangles[t +
                                                    1]]
            vertex3 = self.vertices[self.triangles[t +
                                                    2]]

            p = pygame.Vector3(
                vertex1[0] - vertex2[0],
                vertex1[1] - vertex2[1],
                vertex1[2] - vertex2[2])
            q = pygame.Vector3(
                vertex2[0] - vertex3[0],
                vertex2[1] - vertex3[1],
                vertex2[2] - vertex3[2])
            norm = cross_product(p, q)
            nstart = (0, 0, 0)
            self.normals.append((nstart, nstart +
                                 norm))
```

The initialization method of DisplayNormals.py accepts the vertices and triangles that will be passed when the class is instantiated later. We use the vertices of the triangle to create two vectors that define two sides of the triangle. These vectors, p and q, are then run through the cross-product method to return a normal vector.

The normals that have been calculated are stored in an array as a line segment. In this case, we are drawing the normal from (0, 0, 0), which is the center of the cube, out to the length of the normal:

```python
def draw(self):
    glColor3fv((0, 1, 0))
    glBegin(GL_LINES)
    for i in range(0, len(self.normals)):
        start_point = self.normals[i][0]
        end_point = self.normals[i][1]
        glVertex3fv((start_point[0],
                     start_point[1],
                     start_point[2]))
        glVertex3fv((end_point[0], end_point[1],
                     end_point[2]))
    glEnd()
```

The draw method sets the drawing color to green and then, using the GL_LINES setting, draws the normal as a line segment. Here, start_point and end_point are defined as separate variables to make the code more readable.

5. In Object.py, make the following modifications:

```python
..
from Grid import *
from DisplayNormals import *



class Object:
    def __init__(self, obj_name):
..
elif isinstance(c, Grid):
            c.draw()
        elif isinstance(c, DisplayNormals):
            c.draw()
```

```
        elif isinstance(c, Button):

    ..
```

Here, we are adding the ability for the `Object` update method to handle `DisplayNormal` type components.

6. Finally, we have to update the main program in `Animate.py` to add the normals to a cube. Open this file and make the following modifications:

```
..

from Cube import *
from Grid import *
from DisplayNormals import *
from pygame.locals import *

..

cube = Object("Cube")
cube.add_component(Transform((0, 0, -5)))
cube.add_component(Cube(GL_POLYGON, "images/wall.tif"))
cube.add_component(DisplayNormals(
                cube.get_component(Cube).vertices,
                cube.get_component(Cube).triangles))

cube2 = Object("Cube")
cube2.add_component(Transform((0, 1, -5)))

..
```

This new code will ensure that a `DisplayNormals` component is added to the cube. Note that the `Cube` component, which holds the vertices and triangles, is required to pass the vertices and triangles through to the `DisplayNormals` instance.

7. Running this code will show the animated cube displaying each of its green normals, as shown in *Figure 10.9*:

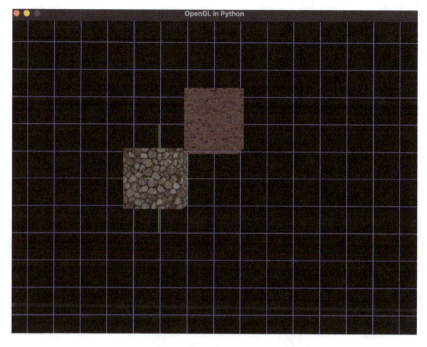

Figure 10.9: A cube displaying its normals

Your turn...

- *Exercise F*:

  - In the previous exercise, we added a class to display the normals on a cube in the project. This code only draws one of the normals. Modify the code so that it calculates and draws both normals.

In this section, we've only touched on the concept of normals by learning to calculate and draw them. Understanding how to do this will help you as you move forward with your learning journey in graphics.

Summary

By now, I am sure you are gaining an appreciation of the importance of vectors, and just how important it is for graphics and game developers to have a firm grasp of them both conceptually and practically. Trust me, this chapter won't be the last time you see them. The mathematical concepts we explored in this chapter might appear to cover the same ground over and over concerning calculating points and vectors, but this should only convince you further how important these methods are.

Lines are far more complex than they first seem. Although most straight geometrical elements are called lines in a general, collective sense, the differences between vectors, lines, line segments, and rays are clear. Each has its place in graphics development in data structures and drawing, though we can apply many of the same mathematical calculations to any of them. Is it any wonder that many of these are stored in the same data structure of `pygame.Vector3`?

We started this chapter by defining the concepts of lines, line segments, rays, and normals. Next, we took this knowledge and applied it to constructing parametric line equations, which were then used in our project to animate a cube so that it moves across the screen. Following this, line drawing was applied to the display of normals to help you visualize the front and back sides of a polygon.

In the next chapter, we will continue using lines, vectors, and rays to further explore the function of normals in drawing, texturing, and lighting 3D objects.

Answers

Exercise A:

end = start + v

end = (3, 6, 7) + (1, 5, 10)

end = (4, 11, 17)

Exercise B:

v = (10, 18, 3) − (1, 9, 5)

v = (9, 9, -2)

t = 0.5

point = (1, 9, 5) + 0.5 * (9, 9, -2)

point = (1, 9, 5) + (4.5, 4.5, -1)

point = (5.5, 13.5, 4)

Exercise C:

start position = (2, 3, 4) − (5, 5, 3) = (-3, -2, 1)

t = 0.75

point = (-3, -2, 1) + 0.75 * (5, 5, 3)

point = (-3, -2, 1) + (3.75, 3.75, 2.25)

point = (0.75, 1.75, 3.25)

Exercise D:

(8, 2, 3) x (1, 2, 3) = (2*3 – 3*2, 3*1 – 8*3, 8*2-2*1)

= (0, -21, 14)

(1, 2, 3) x (8, 2, 3) = (2*3 – 3*2, 3*8 – 1*3, 1*2- 2*8)

= (0, 21, -14)

These two vectors are direct opposites. They are parallel but pointing in opposite directions, which is what you expect from the two normals. You can tell this by looking at the *x*, *y*, and *z* values, each of which has the same values but is signed differently.

Exercise E:

To begin solving this, you need two vectors. These can be calculated from the vertices, as shown in *Figure 10.10*.

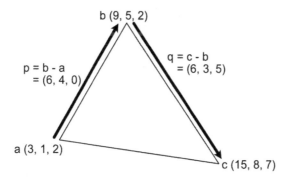

Figure 10.10: Calculating the vectors on a triangle

Then, the cross products are found of *p* x *q* and *q* x *p*, like this:

p x q = (4*5-0*3, 0*6-6*5,6*3-4*6) = (20, -30, -6)

q x p = (3*0-5*4, 5*6-6*0, 6*4-3*6) = (-20, 30, 6)

Notice once again that each normal has the same values, just with different signs.

Exercise F:

To calculate and draw both normals for each of the triangles in the cube, we must add another cross-product calculation and add the extra vector to the normals array, like this:

```
norm1 = cross_product (p, q)
```

```
norm2 = cross_product(q, p)

nstart = (0, 0, 0)

self.normals.append((nstart, nstart + norm1))

self.normals.append((nstart, nstart + norm2))
```

11
Manipulating the Light and Texture of Triangles

We started drawing triangles back in *Chapter 2, Let's Start Drawing*, and added lighting and texturing in *Chapter 5, Let's Light It Up!* Now that we've covered normals in *Chapter 10, Getting Acquainted with Lines, Rays, and Normals*, it's time to put it all together to take a look at how these special vectors are used to manipulate and affect the appearance of triangles.

Normals can belong to vertices and mesh faces. Besides being used to specify which side of a polygon should be rendered, they are also used to calculate how light falls across the surface of a polygon. In this chapter, we will begin by improving on the normal drawing technique discussed in *Chapter 10, Getting Acquainted with Lines, Rays, and Normals*, and use more vector calculations on triangles to find the center of a triangle and draw the normal from that point. This will allow you to draw normals on any mesh in your project, not just cubes. Following this, you will examine how OpenGL uses normals to define the side of a polygon on which a texture will be displayed. In addition, we will explore some special uses of normals in determining whether only one side or both sides of a polygon should be displayed depending on the object that requires rendering. After this, we will delve back into the process of lighting a scene and examine the effect normals have on how light falls across the surface of a polygon.

In this chapter, you will build on your skills in rendering and understanding the nature of 3D model surfaces through the following topics:

- Displaying Mesh Triangle Normals
- Defining Polygon Sides with Normals
- Culling Polygons According to the Normals
- Exploring How Normals Affect Lighting

By the end of this chapter, you will have developed the practical skills to control which sides of a polygon you require to be lit and textured.

Technical requirements

In this chapter, we will continue to work on the project being constructed in Python, PyCharm, Pygame, and PyOpenGL.

The solution files containing the code can be found on GitHub at `https://github.com/PacktPublishing/Mathematics-for-Game-Programming-and-Computer-Graphics/tree/main/Chapter11` in the `Chapter11` folder.

Displaying Mesh Triangle Normals

We will begin this chapter by adding some extra functionality to the normal drawing in *Chapter 10, Getting Acquainted with Lines, Rays, and Normals*. At the time, we restricted the drawing of these normals to the center of the cube being drawn. This technique won't work with more complicated meshes. More ideally, it would work better if each normal for a plane were emitted from the center point of the triangle, which is called the **centroid**. The centroid can be found using the medians of the vectors that make up its sides. Take, for example, the triangle in *Figure 11.1*:

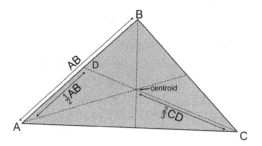

Figure 11.1: The centroid of a triangle

The centroid can be found by finding a point halfway along any of the sides, connecting that point with the opposite corner, and then moving along the vector from the corner to the halfway point by two-thirds. In this example, that means if you find the vector from corner A to B and then travel halfway along it to locate point D, the centroid will be two-thirds of the way along the vector connecting C to D. This is true of the simple calculation made on any of the three sides and their opposite corners.

We will start the practical exercises in this chapter by relocating the starting position of the normals drawn at the end of *Chapter 10, Getting Acquainted with Lines, Rays, and Normals*.

Let's do it...

In this exercise, we will adjust the starting point of the `DisplayNormals` class to allow for the visualization of the normals on any mesh:

1. Begin by making an exact copy of your project folder as it was at the very end of *Chapter 10, Getting Acquainted with Lines, Rays, and Normals*. Name the copied folder `Chapter 11`.

2. Modify the code in `DisplayNormals.py` to calculate and use the median of each triangle like this:

```
...
for t in range(0, len(self.triangles), 3):
    vertex1 = self.vertices[self.triangles[t]]
    vertex2 = self.vertices[self.triangles[t + 1]]
    vertex3 = self.vertices[self.triangles[t + 2]]
    p = pygame.Vector3(vertex1[0] - vertex2[0],
                       vertex1[1] - vertex2[1],
                       vertex1[2] - vertex2[2])
    q = pygame.Vector3(vertex2[0] - vertex3[0],
                       vertex2[1] - vertex3[1],
                       vertex2[2] - vertex3[2])
    norm = cross_product(p, q)
    #find median
    midpoint = vertex3 + q * 0.5
    v = (midpoint - vertex1) * 2/3
    centroid = vertex1 + v
    self.normals.append((centroid, centroid + norm))
```

In this code, the midpoint between `vertex2` and `vertex3` is calculated by adding half of the vector between the points to `vertex3`. A vector between this midpoint and the opposite corner of `vertex1` is calculated and two-thirds of this is used to add to `vertex1`.

> **Important note**
>
> You could use any combination of corners and sides using the same rules to find the centroid. This centroid point is then used as the starting position for the normal.

3. Run the `Animate.py` file that has been copied to the `Chapter 11` folder to see the result. Note the new starting position for the normals, as shown in *Figure 11.2* (with a little embellishment to make it obvious):

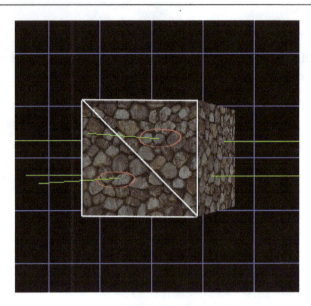

Figure 11.2: A closeup of the normals drawn for each cube triangle

Having made these changes, you will now be able to draw any mesh and display the normals.

4. To have a look at the normals on a mesh loaded from a file, make a copy of `Animate.py` and rename it `ExploreNormals.py`.

5. Download a copy of `planesm.obj` from the `Chapter 11/models` folder on GitHub and place this file in PyCharm in the `Chapter 11/models` folder.

6. From GitHub, download a copy of `ExploreNormals.py` from the `Chapter11/startercode` folder and replace your code with the code in this new file. This code will replace all the cube and grid drawings, as well as the animations with a wireframe view of `planesm.obj`. But don't run it yet as we need to make a few more adjustments to get the code to execute correctly.

7. The plane you are about to draw is facing the camera. This will make it difficult to see the normals. Therefore, we can temporarily rotate and recolor it by changing the code in `Object.py` like this:

```python
def update(self, events = None):
    glPushMatrix()
    for c in self.components:
        if isinstance(c, Transform):
            pos = c.get_position()
            glTranslatef(pos.x, pos.y, pos.z)
            glRotate(45, 0, 1, 0)
```

```
    elif isinstance(c, Mesh3D):
        glColor(1, 1, 1)
        c.draw()
    elif isinstance(c, Grid):
        c.draw()
```

These new lines will rotate the plane by 45 degrees around the *y*- axis and recolor the polygon wireframe to draw in white.

8. One last change to make before running `ExploreNormals.py` is to increase the length of the normals to make them visible. We can do this in `DisplayNormals.py` by adding a multiplier to the end position of the normals, as follows:

```
middle = (vertex3 + q * 0.5)
v = (middle - vertex1) * 2/3
median = vertex1 + v
nstart = median
self.normals.append((nstart, nstart + norm * 10))
```

By multiplying the normal (which has a length of 1) by 10, it will make them go in the same direction but display a longer-drawn line.

9. Now it's time to play `ExploreNormals.py`. When you do, you will see the mesh in white and normals in green, as shown in *Figure 11.3*:

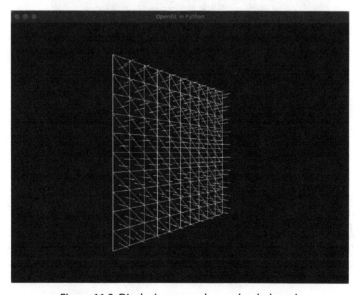

Figure 11.3: Displaying normals on a loaded mesh

Note how each of the normals is drawn from the centroid of each triangle. As was the case with the previous normal calculations in *step 2*, we determined them from the triangle vertices, where the vertices for each triangle were all listed in anticlockwise order.

While the normals included in the `.obj` file face in one direction, mathematically they can face in either direction at 90 degrees to the plane. The effects of this on drawing will now be explored.

Defining polygon sides with normals

So far in this book, we've calculated normals but not explored their many uses. One of these uses is to dictate which side of a polygon is visible. The same plane that you have been using up to this point has had the normals reversed in Autodesk Maya, a 3D modeling program, for the center polygons, as shown in *Figure 11.4*:

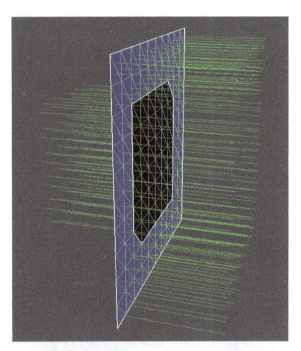

Figure 11.4: A plane with some normals reversed

The black section in the middle of the plane in *Figure 11.4* when rendered would in fact appear as a hole when viewed from one direction and solid from the other. Even though it might look like a hole, that doesn't mean there aren't any polygons covering this area. When this plane is drawn with a cube behind it in Python and OpenGL, the hole is evident, as shown in *Figure 11.5 (a)*:

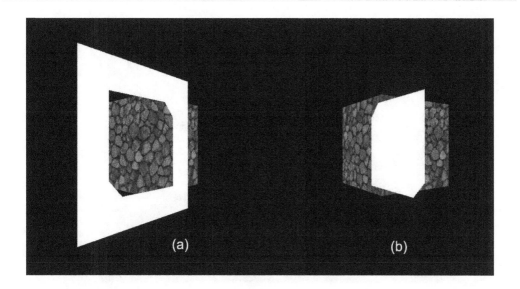

Figure 11.5: A plane with normals reversed viewed from both sides

As can be seen in *Figure 11.5*, whichever polygons have normals on the same side of the camera are opaque. *Figure 11.5 (a)* is a view from the front and *Figure 11.5 (b)* is when the scene is rotated by 180 degrees. What appears as a hole in *Figure 11.5 (a)* can be seen as polygons in *Figure 11.5 (b)*.

In some cases, you'll want to be able to see polygons on just one side, and in others on both sides. In the next section, we will explore these cases.

Working with visible sides of a polygon

Why would we want the option of enabling or disabling the visible sides of polygons? There are two popular cases.

The most common use of a single plane is when it needs to represent something solid. A plane has no thickness mathematically. However, using a box or other 3D shape when only a very thin flat surface is required more than doubles the number of polygons needed to draw the object because a flat plane only has one surface, whereas a cube has six. For example, if you are modeling the interior of a sci-fi corridor, each wall panel, as shown in *Figure 11.6*, only needs to be a single plane with an image on one side:

Figure 11.6: Wall panels represented by a single plane

If the player or viewer is *never* going to see the plane from the other side, then it is a waste of memory and processing effort to try and render it.

The next most common use of a plane to represent something solid is when you need to see a flat plane from both sides, as is the case with leaves. There can be hundreds or even millions of these in a graphics or game environment. The most memory-efficient way of drawing a leaf is with a flat plane displaying the texture, as shown in *Figure 11.7*:

Figure 11.7: A leaf image on a plane

This texture needs to be visible from both sides because you don't want to duplicate the geometry or leaf texture to make it visible on both sides. A plane that is visible on both sides is called a **billboard**. It is used for a variety of purposes, such as faking distant geometry and buildings or for individual leaves and branches on vegetation and grass. Because it is flat, the illusion of 3D that a billboard is attempting to fake will be lost depending on the viewing direction.

Figure 11.8 (a) shows a tree that would be acceptable to a viewer from the ground level, but when viewed from above (*Figure 11.8 (b)*), the illusion is broken and the flat nature of the leaves is noticeable:

Figure 11.8: Branch billboards from different points of view

However, when you want to generate and display an entire forest in 3D, it's in no way practical to accurately model and render individual leaves.

In the next section, you will discover how you can enable and disable the drawing of polygon sides based on their normals.

Culling Polygons According to the Normals

The removal of a polygon from rendering is called **culling**. Culling can occur on an entire polygon or just one side. The side that is removed during the culling process is called the **backface** and it is the opposite side of the polygon to that from which the normal is projected.

In the next practical exercise, we will explore how normals are used to display these types of images.

Let's do it...

So far, the polygons we have been rendering have had textures on both sides, unbeknownst to you. Now it's time to take a look at what's going on both sides:

1. From GitHub, download a copy of `plane2.obj` from the `Chapter 11/models` folder and place it in your project's `model` folder.

2. Make the following changes to `ExploreNormals.py`:

```
from Cube import *
..
objects_3d = []
objects_2d = []
cube = Object("Cube")
cube.add_component(Transform((0, 0, -3)))
cube.add_component(Cube(GL_POLYGON,
                       "images/wall.tif"))
mesh = Object("Plane")
mesh.add_component(Transform((0, 0, -1.5)))
mesh.add_component(LoadMesh(GL_TRIANGLES,
                       "models/plane2.obj"))
objects_3d.append(cube)
objects_3d.append(mesh)
```

The first part of the new code places the textured cube back in the scene. If you've used a different image file to texture your cube, then ensure you use the filename for your texture instead of `wall.tif`.

Other changes have been made to the loading of the mesh to make it draw using `GL_TRIANGLES` instead of `GL_LINE_LOOP`. A new plane model is being used. This model is the one with the apparent hole in it that we saw in *Figure 11.5*.

Also, be sure to remove the `DisplayNormals` component we used on the mesh and cube in the previous exercise.

3. We will now modify the drawing of these 3D objects by allowing them to rotate, and thus allow us to view both sides. To do this, open up the `Object.py` script and make the following changes:

```
class Object:
    def __init__(self, obj_name):
        self.name = obj_name
        self.components = []
```

```
        self.scene_angle = 0
    ..
    def update(self, events = None):
        glPushMatrix()
        for c in self.components:
            if isinstance(c, Transform):
                pos = c.get_position()
                glTranslatef(pos.x, pos.y, pos.z)
                self.scene_angle += 0.5
                glRotate(self.scene_angle, 0, 1, 0)
            elif isinstance(c, Mesh3D):
```

This will keep track of a scene_angle value used to set the viewing angle. By updating the value of this variable in the update method, we can constantly rotate the scene.

4. Run ExploreNormals.py to watch the plane and cube rotating. The plane will look like a square colored entirely in white, as shown in *Figure 11.9*, with the textured cube behind it:

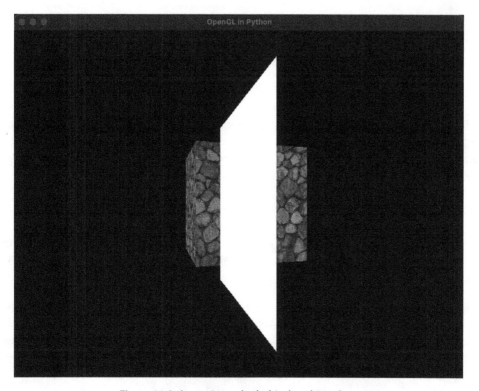

Figure 11.9: A rotating cube behind a white plane

Believe it or not, this plane has the normals in the center reversed like the one in *Figure 11.5*. The reason you can't see the hole or reversal of normals is that by default, OpenGL is drawing both sides of the plane.

5. To only display the side of each polygon from which the normals emanate, edit the `set_3d` method in `ExploreNormals.py` to include the following:

```
..
def set_3d():
    glMatrixMode(GL_PROJECTION)
    ..
    glViewport(0, 0, screen.get_width(),
                screen.get_height())
    glEnable(GL_DEPTH_TEST)
    glEnable(GL_CULL_FACE)
while not done:
    events = pygame.event.get()
..
```

The `GL_CULL_FACE` directive tells OpenGL to ignore drawing the backface of polygons.

6. Run `ExploreNormals.py` again to see the plane drawing with the normals reversed in the middle, as shown in *Figure 11.5*.

The `GL_CULL_FACE` setting is OpenGL's command to prevent the reverse side of polygons from being drawn. This process is called **backface culling**. This dramatically increases performance as it cuts down how many polygons in a scene get rendered. It can be turned on with `glEnable(GL_CULL_FACE)` and turned off with `glDisable(GL_CULL_FACE)`, so the programmer can control which meshes draw on both sides and which don't. We will demonstrate this in the next exercise.

Let's do it...

As previously discussed in the *Working with visible sides of a polygon* section, one type of mesh that would require both sides to be drawn is for leaves. In this exercise, we will add this object, and you will learn how to turn backface culling on and off:

1. Before we can show both sides of a mesh being drawn with a texture, the `LoadMesh` class needs to be updated to accommodate the addition of a texture. Much of this code should be familiar to you from the `Mesh3D` class. Open `LoadMesh.py` and make the following changes:

```
class LoadMesh(Mesh3D):
    def __init__(self, draw_type, model_filename,
                    texture_file=""):
```

```
    self.vertices, self.uvs, self.triangles =
                    self.load_drawing(model_filename)
    self.texture_file = texture_file
    self.draw_type = draw_type
    if self.texture_file != "":
        self.texture =
            pygame.image.load(texture_file)
        self.texID = glGenTextures(1)
        textureData = pygame.image.tostring(
                    self.texture, "RGB", 1)
        width = self.texture.get_width()
        height = self.texture.get_height()
        glBindTexture(GL_TEXTURE_2D, self.texID)
        glTexParameteri(GL_TEXTURE_2D,
                    GL_TEXTURE_MIN_FILTER,
                    GL_LINEAR)
        glTexImage2D(GL_TEXTURE_2D, 0, 3, width,
            height,
            0, GL_RGB, GL_UNSIGNED_BYTE,
            textureData)
```

The first changes to the code made in the __init__ method test to see whether a texture filename has been passed through. If it has, then the same code used in Mesh3D.py to load in a texture and initialize the required parameters is included.

The next changes occur in the draw method of LoadMesh.py:

```
def draw(self):
    if self.texture_file != "":
        glEnable(GL_TEXTURE_2D)
        glTexEnvf(GL_TEXTURE_ENV,
                GL_TEXTURE_ENV_MODE,
                GL_DECAL)
        glBindTexture(GL_TEXTURE_2D, self.texID)
    for t in range(0, len(self.triangles), 3):
        glBegin(self.draw_type)
        if self.texture_file != "":
            glTexCoord2fv(
                self.uvs[self.triangles[t]])
```

```
      glVertex3fv(
          self.vertices[self.triangles[t]])
      if self.texture_file != "":
          glTexCoord2fv(
              self.uvs[self.triangles[t + 1]])
      glVertex3fv(self.vertices[self.triangles[t
              + 1]])
      if self.texture_file != "":
          glTexCoord2fv(
              self.uvs[self.triangles[t + 2]])
      glVertex3fv(self.vertices[self.triangles[t
              + 2]])
      glEnd()
  if self.texture_file != "":
      glDisable(GL_TEXTURE_2D)
```

Within the draw method, the texture drawing is enabled if there is a texture file. If there's no texture filename given to the class, the mesh draws as it did before, without a texture and in white.

As we are now using a texture, uvs values must be included with the drawing of each vertex, and so checks for the texture file are performed again before the uvs values are used:

```
def load_drawing(self, filename):
    vertices = []
    uvs = []
    triangles = []
    with open(filename) as fp:
        line = fp.readline()
        while line:
            if line[:2] == "v ":
                ..
            if line[:2] == "vt":
                vx, vy = [float(value) for value
                            in line[3:].split()]
                uvs.append((vx, vy))
            if line[:2] == "f ":
                ..
            line = fp.readline()
    return vertices, uvs, triangles
```

Lastly, the `load_drawing` method is expanded to read in the lines from the OBJ file that contain the `uvs` values. This is a nice feature as it saves us from typing them in manually.

2. `ExploreNormals.py` can now be changed to draw two planes and a cube. We will draw both the previous plane without a hole and the one with the hole. The plane without the hole will be textured. Be sure to use your own texture file in place of the one I have used:

```
objects_2d = []
cube = Object("Cube")
cube.add_component(Transform((0, 0, -3)))
cube.add_component(Cube(GL_POLYGON,
                        "images/wall.tif"))
mesh = Object("Plane")
mesh.add_component(Transform((-1.5, 0, -3.5)))
mesh.add_component(LoadMesh(GL_TRIANGLES,
                           "models/plane2.obj"))
leaf = Object("Leaf")
leaf.add_component(Transform((0.5, 0, -1.5)))
leaf.add_component(LoadMesh(GL_TRIANGLES,
                           "models/plane.obj",
                           "images/brick.tif"))
objects_3d.append(cube)
objects_3d.append(mesh)
objects_3d.append(leaf)
clock = pygame.time.Clock()
```

Here, you can see the drawing will have a cube and two planes. Be sure to change the positions of each via its transform values to ensure they are all visible in the drawing.

Finally, comment out or remove the GL_CULL_FACE enabling:

```
def set_3d():
    glMatrixMode(GL_PROJECTION)
    ..
    glEnable(GL_DEPTH_TEST)
    #glEnable(GL_CULL_FACE)
```

Run `ExploreNormals.py` and note how the backfaces of all drawn polygons are visible, as shown in *Figure 11.10*:

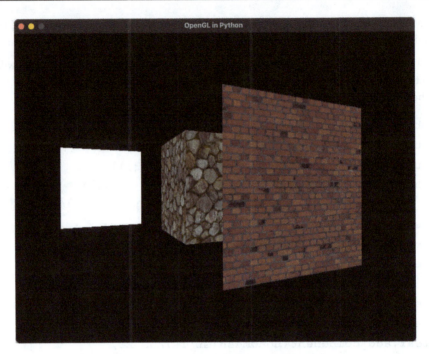

Figure 11.10: Drawing without culling backface polygons

3.   Edit LoadMesh.py and modify the code to accept a parameter to enable back_face_cull like this:

```
class LoadMesh(Mesh3D):
    def __init__(self, draw_type, model_filename,
        texture_file="", back_face_cull=False):
        ..
        self.draw_type = draw_type
        self.back_face_cull = back_face_cull
```

First, an optional parameter for the LoadMesh class is passed through from where the class is being instantiated. Then, in the draw method, the same parameter is used to turn backface culling on and off:

```
def draw(self):
    if self.back_face_cull:
        glEnable(GL_CULL_FACE)
    if self.texture_file != "":
        glEnable(GL_TEXTURE_2D)
        ..
```

```
          if self.texture_file != "":
              glDisable(GL_TEXTURE_2D)
          if self.back_face_cull:
              glDisable(GL_CULL_FACE)
```

4. Lastly, we will have the plane with the hole in it, using backface culling. To do this, modify its creation code in ExploreNormals.py, as follows:

```
mesh = Object("Plane")
mesh.add_component(Transform((-1.5, 0, -3.5)))
mesh.add_component(LoadMesh(GL_TRIANGLES,
                           "models/plane2.obj",
                           back_face_cull=True))
```

5. Now, when you run ExploreNormals.py, the plane containing the hole will draw only the sides of the polygons with the normals. The other plane will look the same on both sides, as shown in *Figure 11.11*:

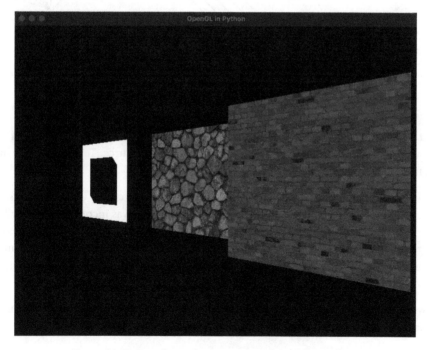

Figure 11.11: Drawing only one mesh with backface culling

In this section, you have learned how to pull in the normals from an OBJ mesh file and use them to control the sides of a polygon that are rendered. This is a valuable skill in learning how to optimize the drawing speed of 3D scenes by eliminating the sides that don't need to be processed.

Normals are not only used to determine the front and backfaces of polygons but are also essential in calculating lighting effects. A preliminary exploration of normals and lights will be presented next.

Exploring How Normals Affect Lighting

Lighting is a complex topic in graphics. So far, we've applied very simple ambient and diffuse lighting to a cube in *Chapter 5, Let's Light It Up!* At the time, we discussed the lighting model of specular reflection but didn't practically apply it as it requires the use of normals. For your convenience, *Figure 5.2 of Chapter 5, Let's Light It Up!*, is repeated here as *Figure 11.12*:

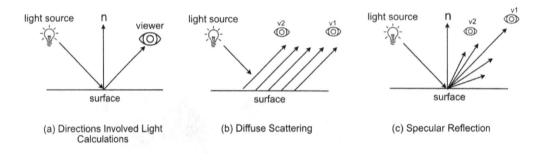

Figure 11.12: The components of light that make up a final render

For ambient and diffuse lighting, the normal is not used and as such makes the surface of meshes appear flat with very little indication of the direction of the light source. To add specular lighting though, we need to know the normal.

Although we didn't specify the normals for the cube in the light exercise in *Chapter 5, Let's Light It Up!*, we were still able to get a lighting effect. However, this lighting was incorrect, as you will see in the next exercise. When OpenGL doesn't receive any information about normals, it calculates its own by assuming the vertices for triangles are given to it in counter-clockwise order. However, unless we specify the order of vertices ourselves, we can't be sure they are in counter-clockwise order, especially when an external file format such as OBJ is being used.

In the following exercise, we will take a quick look at the effect of lighting and adding normals, while leaving the hardcore lighting applications and discussion for *Chapter 18, Customizing the Render Pipeline*.

Let's do it...

In this practical exercise, we are going to add light to a scene and compare OpenGL's normal calculations to our own:

1. Create a new Python script called `NormalLights.py`.

2. Copy the code from `ExploreNormals.py` into `NormalLights.py`.

3. Download a copy of `cube.obj` in the `models` folder from `Chapter 11` on GitHub. This file contains normal values for a cube model.

4. Place `cube.obj` into your `models` folder for the current project.

5. Modify `NormalLights.py` to turn on some lights by adding the following code to the `set_3d()` method:

```
def set_3d():
    glMatrixMode(GL_PROJECTION)
    glLoadIdentity()
    gluPerspective(60, (screen_width / screen_height),
                   0.1,
                   100.0)
    glMatrixMode(GL_MODELVIEW)
    glLoadIdentity()
    glViewport(0, 0, screen.get_width(),
               screen.get_height())
    glEnable(GL_DEPTH_TEST)
    glEnable(GL_LIGHTING)
    glLight(GL_LIGHT0, GL_POSITION, (5, 5, 5, 0))
    glLightfv(GL_LIGHT0, GL_AMBIENT, (1, 0, 1, 1))
    glLightfv(GL_LIGHT0, GL_DIFFUSE, (1, 1, 0, 1))
    glLightfv(GL_LIGHT0, GL_SPECULAR, (0, 1, 0, 1))
    glEnable(GL_LIGHT0)
```

This code is the same code we used in *Chapter 5, Let's Light It Up!* It enables lighting, positions a light at (5, 5, 5, 0), and then specifies the colors for the ambient, diffuse, and specular lights, before enabling the only light in the scene.

6. Reduce the number of objects being drawn in the scene by having just one that draws a cube from the `cube.obj` file by modifying `NormalLights.py` like this:

```
..
objects_3d = []
```

```
objects_2d = []
cube = Object("Cube")
cube.add_component(Transform((0, 0, -3)))
cube.add_component(LoadMesh(GL_TRIANGLES,
                            "models/cube.obj"))
objects_3d.append(cube)
clock = pygame.time.Clock()
fps = 30
..
```

7. Press **play** to see a single cube rotating in the scene. Take note of how it changes color. It seems to randomly change from yellow to purple without giving any indication of the direction of the light source that we placed at (5, 5, 5), as shown in *Figure 11.13*:

Figure 11.13: The components of light that make up a final render

Whatever normal values OpenGL is calculating are wrong. The cube should appear to be lit from a single direction. In addition, OpenGL is only using one normal per triangle face instead of a normal for each vertex.

8. We will now modify LoadMesh.py to load the normals out of the OBJ file, as follows:

```
class LoadMesh(Mesh3D):
    def __init__(self, draw_type, model_filename,
                 texture_file="",
                 back_face_cull=False):
        self.vertices, self.uvs, self.normals,
        self.normal_ind, self.triangles =
                 self.load_drawing(model_filename)
        self.texture_file = texture_file
```

In this code, we are going to get a set of normals and a set of normal indices returned from the loading of the mesh. The indices for the normals act in the same way as the triangles do for the vertices to make sure they are used in the correct order.

9. Continue to edit LoadMesh.py to load in the normals and normal indices from the OBJ file like this:

```python
    ..
    def load_drawing(self, filename):
        vertices = []
        uvs = []
        normals = []
        normal_ind = []
        triangles = []
        with open(filename) as fp:
            line = fp.readline()
            while line:
                if line[:2] == "v ":
                    vx, vy, vz = [float(value) for value
                                in line[2:].split()]
                    vertices.append((vx, vy, vz))
                if line[:2] == "vn":
                    vx, vy, vz = [float(value) for value
                                in line[3:].split()]
                    normals.append((vx, vy, vz))
                if line[:2] == "vt":

                    ..

                if line[:2] == "f ":
                    t1, t2, t3 = [value for value in
                                line[2:].split()]

                    ...

                    triangles.append([int(value) for value
                                in
                                t3.split('/')][0] -
                                1)
                    normal_ind.append([int(value) for
                                value in
                                t1.split('/')][2] -
```

```
                                                 1)
                    normal_ind.append([int(value) for
                                       value in
                                       t2.split('/')][2] -
                                       1)
                    normal_ind.append([int(value) for
                                       value in
                                       t3.split('/')][2] -
                                       1)
            line = fp.readline()
    return vertices, uvs, normals, normal_ind,
        triangles
```

As you can see, reading in the normals and normal indices from the file is done in the same manner as getting vertices, uvs, and triangles. A line beginning with vn indicates it is a normal.

10. Last, we need to use these normals when drawing the cube. Therefore, we modify the draw() method of LoadMesh, as follows:

```
    ..
    def draw(self):
        ..
        for t in range(0, len(self.triangles), 3):
            glBegin(self.draw_type)
            if self.texture_file != "":
                glTexCoord2fv(self.uvs[self.triangles[t]])
            glNormal3fv(self.normals[self.normal_ind[t]])
            glVertex3fv(self.vertices[self.triangles[t]])
            if self.texture_file != "":
                glTexCoord2fv(self.uvs[self.triangles[t +
                                                      1]])
            glNormal3fv(self.normals[self.normal_ind[t +
                                                     1]])
            glVertex3fv(self.vertices[self.triangles[t +
                                                     1]])
            if self.texture_file != "":
                glTexCoord2fv(self.uvs[self.triangles[t +
                                                      2]])
```

```
glNormal3fv(self.normals[self.normal_ind[t +
                                         2]])
glVertex3fv(self.vertices[self.triangles[t +
                                         2]])

glEnd()
```

The code includes a call to the `glNormal3fv()` method that is similar to a `glVertex3fv()` call in which it takes a vector with *x*, *y*, and *z* values to specify the normal. Note that `glNormal3fv` is called before `glVertex3fv`.

11. Run this now and you will see a cube that is lit correctly with a single light source, as shown in *Figure 11.14*:

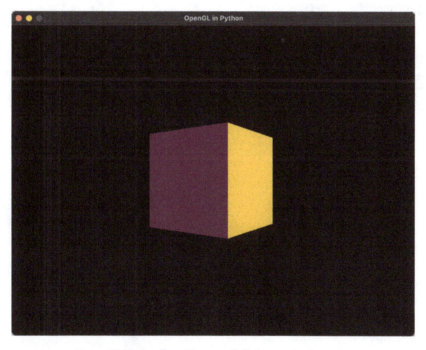

Figure 11.14: A cube with normals lit from a single source

This exercise has illustrated the use of normals and their importance in lighting situations. It always pays to critically evaluate the output of any program, especially graphics. Don't assume something must be correct even if it appears wrong and the program runs.

Summary

In this chapter, we have explored the importance of normals, not only for determining the front and backfaces of a polygon but also for controlling lighting. We started by examining how a proper normal could be drawn and calculated it for each triangle face before using backface culling to investigate how both or a single side of a polygon could be drawn. Following this, we added to our project the ability to load normals calculated in a modeling package out of an OBJ file and into OpenGL. We illustrated just how important it is, in rendering, to ensure you specify the correct values instead of making certain assumptions if the output doesn't look quite right.

As we progress through the book, we will encounter more and more uses of normals, including their use in calculating collisions, as well as creating special texturing and lighting effects. However, I am sure you are beginning to gain an appreciation of their importance.

By now, you should be comfortable with the concept of normals, how to calculate them, and how to load them from a file. They might just be a simple vector attached to polygon faces and vertices but they are an incredibly powerful concept.

In the next chapter, we will get things moving, literally, by expanding our work on transformations and the transformations class in our project. In it, you will discover how to freely move, scale, and rotate objects, as well as investigating the powerful mathematical concept of matrices that underpins all calculations that occur in graphics engines.

Part 3 – Essential Transformations

In this part, you will learn how to manipulate graphical elements primarily through translation, rotation, and scaling. This content makes use of the knowledge obtained in *Part 2*, *Essential Trigonometry*, by applying vector and triangle mathematics to solve the problem of moving, scaling, and rotating objects in a 3D environment. Of most importance to your success as a graphics programmer is the thorough examination of matrices and quaternions that will explain why graphics and game engines always work with 4D values. In addition, the perplexing set of coordinate spaces that are a large part of transforming a model's 3D vertices into pixels on the screen will be demystified and built into the project.

In this part, we cover the following chapters:

- *Chapter 12, Mastering Affine Transformations*
- *Chapter 13, Understanding the Importance of Matrices*
- *Chapter 14, Working with Coordinate Spaces*
- *Chapter 15, Navigating the View Space*
- *Chapter 16, Rotating with Quaternions*

12

Mastering Affine Transformations

Throughout the book so far, you'll have gained an appreciation for the variety of methods used to move, rotate, and scale vectors and points. To move a mesh from one location in space to another requires each vertex in that mesh to be moved, and then the mesh is redrawn. This movement (formally called a **translation**) is just one of a set of special point and vector manipulator methods called **affine transformations**.

Affine transformations are important in computer graphics primarily, as they allow for the manipulation of a set of vertices without losing the integrity of the form. By this, I mean that any lines and planes in a mesh retain their relative parallelism and ratios. This might sound a little abstract, so let's illustrate it with an example. Consider the diagram in *Figure 12.1*:

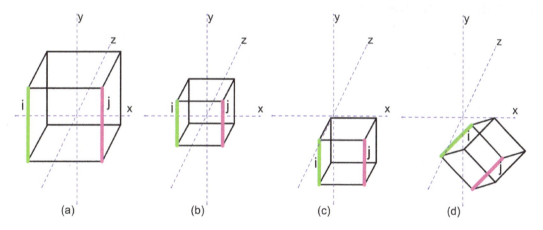

Figure 12.1: An affine transformation of a cube

Figure 12.1 shows an original cube in (**a**) made up of six sides and eight vertices. Through a series of **scaling** (**b**), **translations** (**c**), and **rotations** (**d**), it ends up as the rectangular prism in (**d**). Notice how each of the sides and edges of the cube that were parallel before the transformation is still parallel after? This is an affine transformation. Not only is parallelism maintained but the ratio between the sizes of the parallel planes or edges also is. In this case, the ratio of the lengths of *i* and *j* in (**a**) will be the same as the ratio of the lengths of *i* and *j* in (**b**). If *i* and *j* are both a length of 2 in (**a**), this gives a ratio of 1. So, if *i* is 0.5 in (**b**), then *j* will also be a length of 0.5.

In this chapter, we will discuss each of the affine transformations and explore the mathematics that makes them possible, as well as the advantages they bring to computer graphics. Each operation will be discussed separately as we work to build them into our Python project. The topics we will cover will include:

- Translating points in 3D
- Scaling points in *x*, *y*, and *z*
- Rotating points around a pivot
- Exploring transformation orders
- Shearing and reflections

By the end of this chapter, you will have explored the mathematics of transformations and be confident in using them to reposition graphics objects in a virtual environment. This will give you the ability to set up complex scenes and create animations.

Technical requirements

In this chapter, we will be using Python, PyCharm, and Pygame, as used in previous chapters.

Before you begin coding, create a new `Chapter_12` folder in the PyCharm project for the contents of this chapter.

The solution files containing the code can be found on GitHub at `https://github.com/PacktPublishing/Mathematics-for-Game-Programming-and-Computer-Graphics/tree/main/Chapter12`.

Translating points in 3D

We first encountered OpenGL's translations and rotations back in *Chapter 4, Graphics and Game Engine Components*, and then began building a `Transform` class for our Python project in *Chapter 6, Updating and Drawing the Graphics Environment*. In this chapter, we will continue working on this class to provide all the functionality of affine transformations, beginning with that of translation.

Whichever affine transformation you are applying, the rule is that the operation is applied to every point to which you want to apply the transformation. In the case of the vertices of a cube, you may have six vertices, as shown in *Figure 12.2*. This cube is centered around the origin:

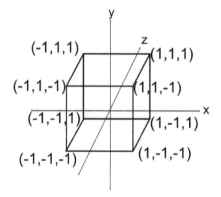

Figure 12.2: A cube with each vertex coordinate displayed

To translate the cube—that is, to move it to another location—we must perform the same operation on each of the vertices such that they maintain the integrity of the cube. That means that it must still look like a cube after it has been moved. The mathematics of a translation is simple. The coordinates of each point have a value added or subtracted from it. More formally, the equation can be written like this:

$$P_{(x, y, z)} = T_{(x, y, z)} + Q_{(x, y, z)}$$

Here, P is the resulting point, T is the translation amount, and Q is the original point.

> **Note**
>
> If the original point is in 3D, then the translation must include values to add to *x*, *y*, and *z*, which will then result in a final point that is also in 3D.

In *Chapter 7, Interactions with the Keyboard and Mouse for Dynamic Graphics Programs*, we added a method that performs this operation into the `Transform` class, thus we get the following code:

```
def move(self, amount: pygame.Vector3):
    self.position = pygame.Vector3(self.position.x +
                                   amount.x, self.position.y
                                   + amount.y,
                                   self.position.z +
                                   amount.z)
```

This code takes a movement amount—that is, a `Vector3` value—and adds an x amount to the x position, a y amount to the y position, and a z amount to the z position to calculate a movement.

If the cube shown in *Figure 12.2* were translated by (3, 2, 1), then the new vertices would be positioned as shown in *Figure 12.3*:

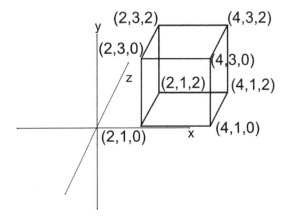

Figure 12.3: A translated cube and vertices

Note how all the *x* coordinates have had 3 added, the *y* coordinates have 2 added, and the *z* coordinates have 1 added. In addition, the cube is still the same-sized cube that we started with, just in a new location.

Your turn...

Exercise A. What is the resulting point if point (3, 4) is translated by (1, 5)?

Exercise B. What is the resulting point if point (2, 3) is translated by (-3, 2)?

Exercise C. What is the resulting point if point (5, 4, 2) is translated by (0, -4, 6)?

As you've seen in this section, translating is a simple addition operation that moves points from one location to another in space. When a point has its coordinates multiplied, then the resulting object constructed from the points is scaled, which we will discuss next.

Scaling points with x, y, and z

It might seem a strange proposition to scale a single point if you think about it, as a point has no size—it's just a location in space. So, what happens if you try to scale it through the affine transformation of scaling? Well, scaling is an operation performed by the multiplication of each of the point coordinates. Take, for example, the point (2, 4, 6)—if this is scaled by 0.5 (in other words, halved), the resulting point is (1, 2, 3). In this case, what has happened to the point is that it has been moved.

The formal mathematics for scaling is:

$$P_{(x, y, z)} = S \times Q_{(x, y, z)}$$

Here, the x, y, and z coordinates of the resulting point P are the point Q's individual coordinates multiplied by S. Let's consider again the cube from *Figure 12.2*. The result of multiplying each of the cube's vertices by 0.5 will result in the cube shown in *Figure 12.4*:

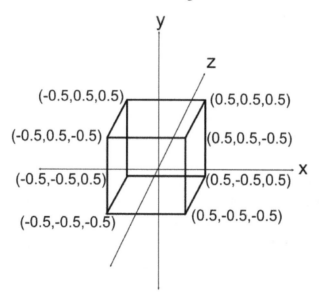

Figure 12.4: A scaled cube

In this case, note that each of the vertices has moved closer to the origin, but the cube shape they form is still a cube. It's just half the size of the original cube.

It's much easier to examine the process of scaling in practice, so let's add this functionality to our Python project.

Let's do it...

In this practical exercise, we are going to add scaling functionality to the `Transform` class and then investigate how scaling objects works:

1. Make a new `Chapter_12` folder in PyCharm and copy the contents of the `Chapter_11` folder into it.

2. Create a new Python script called `ScalingObjects.py` and add a copy of the code from `Vectors.py` in `Chapter_9` to it.

3. When you run `ScalingObjects.py`, you will notice a familiar scene of a blue grid with two cubes on top of each other. The whole scene will, however, be rotating, thanks to some leftover code in `Object.py`. Let's remove this as it will no longer be needed, and while we are modifying the code, we will add new lines to work with scaling.

   Open `Object.py` and modify the update method, like this:

   ```
   def update(self, events = None):
       glPushMatrix()
       for c in self.components:
           if isinstance(c, Transform):
               pos = c.get_position()
               scale = c.get_scale()
               glTranslatef(pos.x, pos.y, pos.z)
               glScalef(scale.x, scale.y, scale.z)
           elif isinstance(c, Mesh3D):
       ..
   ```

   This code gets the object's scale from the `Transform` class, and then uses the `glScalef()` OpenGL method to set the scales of the *x*, *y*, and *z* coordinates separately.

4. To accommodate the use of this scaling when the object is being drawn, we must update the `Transform` class thus:

   ```
   class Transform:
       def __init__(self, position=pygame.Vector3(0, 0,
                       0), scale=pygame.Vector3(1, 1, 1)):
           self.position = pygame.Vector3(position)
           self.scale = pygame.Vector3(scale)

       ..

       def get_scale(self):
           return self.scale
   ```

```
    def set_scale(self, amount: pygame.Vector3):
        self.scale = amount
```

This code adds scale as a property of the `Transform` class as well as setting default values for the position and scale in the constructor. Note that a scale of 1 will do nothing to the object. At the end of the class, two new methods, `get_scale()` and `set_scale()`, are added to allow the scale to be set and retrieved as required.

5. Finally, the code for `ScalingObjects.py` requires updating. We will only want one cube for now, so any other objects can be removed thus:

```
..
objects_3d = []
objects_2d = []

cube = Object("Cube")
cube.add_component(Transform((0, 0, -5)))
cube.add_component(Cube(GL_POLYGON,
                "images/wall.tif"))

objects_3d.append(cube)

grid = Object("Grid")
grid.add_component(Transform((0, 0, -5)))
..
```

In the code you copied from `Vectors.py` in *step 2*, you will notice there is a second cube called `cube2` added. All references to this can be removed to leave just one object and the grid.

6. In the same `ScalingObjects.py` code, references to `cube2` can be removed before the main game loop, the spacebar controls removed, and the up- and down-arrow keys programmed to scale the cube up and down, like this:

```
..
glViewport(0, 0, screen.get_width(),
           screen.get_height())
glEnable(GL_DEPTH_TEST)



trans: Transform = cube.get_component(Transform)
```

```
while not done:
    events = pygame.event.get()
    for event in events:
        if event.type == pygame.QUIT:
            done = True

    keys = pygame.key.get_pressed()
    if keys[pygame.K_UP]:
        trans.set_scale(pygame.Vector3(2, 2, 2))
    if keys[pygame.K_DOWN]:
        trans.set_scale(pygame.Vector3(0.5, 0.5, 0.5))

    glPushMatrix()
    ..
```

7. When you run this, you will be able to scale the original cube by half by pressing the down-arrow key and scale it by two using the up-arrow key, as shown in *Figure 12.5*:

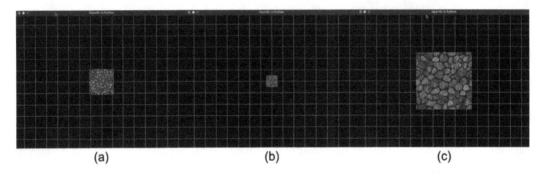

(a)   (b)   (c)

Figure 12.5: Scaling a cube

Notice that when you press the arrow keys, the cube isn't continually rescaled, such that it won't get larger and larger or smaller and smaller, because it is always using the cube's original set of vertices from which the cube is scaled each time. I'll leave this particular functionality for you as the next challenge.

Your turn...

Exercise D. Modify the current state of your Python project such that continually pressing the up-arrow key will compound the effect of scaling up by two and continually pressing the down-arrow key will compound the scaling-down effect.

> **Hint**
>
> Take note in the `Transform` class of how the position is constantly updated. That is how we achieved previous animation effects of the cube moving across the screen. Currently, the code just sets the scale rather than updating it.

In this section, we have investigated scaling and how it is different from translation, in which it uses multiplication as the main operation. You may scale uniformly in each direction or apply different scales for elongated and squashed effects. Even though you may be able to squash a cube into a rectangular prism, it is still considered an affine transformation as the edges remain parallel and the comparative lengths of edges keep the same ratio.

Translation and scaling would have to be the easiest of the transformations to work with. In the next section, we will investigate rotation. You will soon see that it is far more complex.

Rotating points around a pivot

Right now, you are probably feeling pretty comfortable with affine transformations. There's no doubt that scaling and translation are simple concepts. But now, we move on to rotations. Hold on to your hat because the mathematics is about to go up a few notches in complexity.

Just as translation and scaling in 3D work with each of the x, y, and z axes, so too does rotation. An object can rotate around its x, y, or z axis. These rotations are illustrated in *Figure 12.6*:

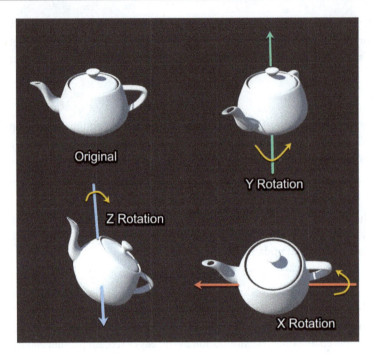

Figure 12.6: A teapot rotated around each axis

What makes the calculations for each of these rotations more difficult than scaling and translation is that while the x, y, and z values applied in scaling and translation only affect their coordinate counterparts (for example, x affects *x*, y affects *y*, and likewise), with rotations, to rotate around one axis, the other two axes are involved. The following equations use what are called **Euler angles**. These were introduced by the mathematician **Leonhard Euler**. He devised them to describe the three rotations around each of the x, y, and z angles. Here are the official rotation equations:

- To perform a rotation around the *x* axis (otherwise called a *pitch*):

$$P_x = Q_x$$

$$P_y = Q_y \times \cos(\theta) + Q_z \times \sin(\theta)$$

$$P_z = Q_y \times -\sin(\theta) + Q_z \times \cos(\theta)$$

- To perform a rotation around the *y* axis (otherwise called a *yaw*):

$$P_x = Q_x \times \cos(\theta) - Q_z \times \sin(\theta)$$

$$P_y = Q_y$$

$$P_z = Q_x \times \sin(\theta) + Q_z \times \cos(\theta)$$

- To perform a rotation around the *z* axis (otherwise called a *roll*):

$$P_x = Q_x \times \cos(\theta) + Q_y \times \sin(\theta)$$

$$P_y = Q_x \times -\sin(\theta) + Q_y \times \cos(\theta)$$

$$P_z = Q_z$$

Notice the difference between translation and scaling? Whereas the other transformations involved the same equations for each x, y, and z value, with rotations, each value now has its own equation, but each of these equations is completely different depending on the axis of rotation.

If you are interested in the derivation of these equations, I would strongly encourage you to take a look at https://www.khanacademy.org/computing/computer-programming/programming-games-visualizations/programming-3d-shapes/a/rotating-3d-shapes.

Luckily for us, OpenGL simplifies the process of rotating objects with its glRotated() method that we used in *Chapter 4, Graphics and Game Engine Components*. However, a fundamental understanding of how these rotations work will assist you in evaluating the results you see on the screen.

We'll now work on adding rotational functionality to the Transform class of your project so that you can get some practice.

Let's do it...

For this practical exercise, we will add the ability to rotate an object by placing the relevant code into the Transform class. Follow the next steps:

1. Open Transform.py and make the following additions to work with a rotation axis and a rotation angle:

```
class Transform:
    def __init__(self, position=pygame.Vector3(0, 0,
                                                0),
                 scale=pygame.Vector3(1, 1, 1)):
        self.position = pygame.Vector3(position)
        self.scale = pygame.Vector3(scale)
        self.rotation_angle = 0
        self.rotation_axis = pygame.Vector3(0, 1, 0)

    ..

    def get_rotation_angle(self):
        return self.rotation_angle
```

```
def get_rotation_axis(self):
    return self.rotation_axis

def set_rotation_axis(self, amount: pygame.Vector3):
    self.rotation_axis = amount

def update_rotation_angle(self, amount):
    self.rotation_angle += amount
```

Here, we are adding properties to store a rotation angle and a rotation axis. These can be set and retrieved using the `get_rotation_angle()`, `get_rotation_axis()`, `set_rotation_axis()`, and `update_rotation_angle()` methods.

2. Next, the `update()` method in `Object.py` needs to be modified to work with these rotation values, thus:

```
def update(self, events = None):
    glPushMatrix()
    for c in self.components:
        if isinstance(c, Transform):
            pos = c.get_position()
            scale = c.get_scale()
            rot_angle = c.get_rotation_angle()
            rot_axis = c.get_rotation_axis()
            glTranslatef(pos.x, pos.y, pos.z)
            glScalef(scale.x, scale.y, scale.z)
            glRotated(rot_angle, rot_axis.x,
                    rot_axis.y,
                    rot_axis.z)
        elif isinstance(c, Mesh3D):
```

In this code, after we get the rotation angle and axis from the transform component, it is used in a `glRotated()` method to rotate the object by `rot_angle` degrees around the axis defined by `rot_axis`.

3. To get the object to rotate with the left- and right-arrow keys, it's simply a matter of adding the following code in the main loop of `ScalingObjects.py`, like this:

```
..
if keys[pygame.K_DOWN]:
    trans.update_scale(pygame.Vector3(0.5, 0.5, 0.5))
```

```
if keys[pygame.K_LEFT]:
    trans.update_rotation_angle(5)
if keys[pygame.K_RIGHT]:
    trans.update_rotation_angle(-5)

glPushMatrix()
..
```

If you'd prefer a faster rotation when the program runs, simply change the amount of angle rotation from 5 to a higher number. Note the use of the minus (-) sign makes the object rotate in the opposite direction. By default, we have programmed the object to rotate around its vertical or up (y) axis, though you can experiment with others.

4. Run the project. You will be able to spin the cube by holding down the left- or right-arrow keys. You can still scale with the up- and down-arrow keys, and then continue rotating. The cube will rotate around its axis, as shown in *Figure 12.7*:

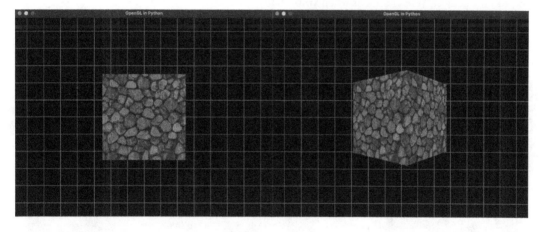

Figure 12.7: The cube with a rotation of 0 on the left and a rotation of 45 on the right

In this section, you've discovered how to add simple OpenGL rotations controlled by the arrow keys into the `Transform` class to animate the revolution of an object. But have you stopped to consider the order we typed in the transformation commands? What if they were in a different order? Would it matter? These questions will be answered in the next section.

Exploring transformation orders

Assuming you typed in your code in the same order that I did, then you'd have your transformations listed like this within the `Object.py` code:

```
glTranslatef(pos.x, pos.y, pos.z)
glScalef(scale.x, scale.y, scale.z)
glRotated(rot_angle, rot_axis.x, rot_axis.y, rot_axis.z)
```

But did you wonder why they were in that order? Why don't you place the `glRotated()` line first in this list, like this:

```
glRotated(rot_angle, rot_axis.x, rot_axis.y, rot_axis.z)
glTranslatef(pos.x, pos.y, pos.z)
glScalef(scale.x, scale.y, scale.z)
```

Now, run the project. What happens when you rotate the cube?

It goes off the screen, right? However, it appears to be slightly rotating. Hold down the right-arrow key for a while. The cube will go off the right-hand side of the window, and then eventually come back from the left-hand side, as illustrated in *Figure 12.8*:

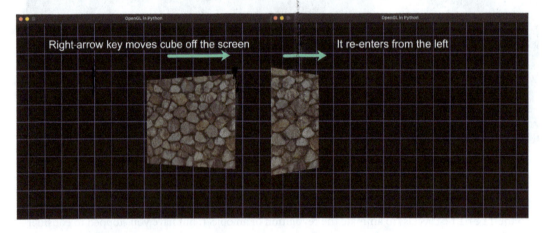

Figure 12.8: The cube spinning around the viewer's head

What's happening, in this case, is that the cube is rotating around the virtual world's (0, 0, 0) point. When it goes off the screen, you could imagine it spinning around the back of your head, and then back onto the screen as it re-enters from the opposite direction.

Why is the cube behaving so differently? It has to do with the way that OpenGL stacks transformations before applying them to drawn objects.

If you were to only perform the `glRotated()` method without the scale and translation, the cube would be drawn at the origin or (0, 0, 0), and then it would rotate around that spot. In fact, all rotations are applied around the origin. It's obvious when the only transformation you do is rotation. Basically, every time you apply a transformation, you are altering the drawing frame of reference. Take a look at *Figure 12.9*:

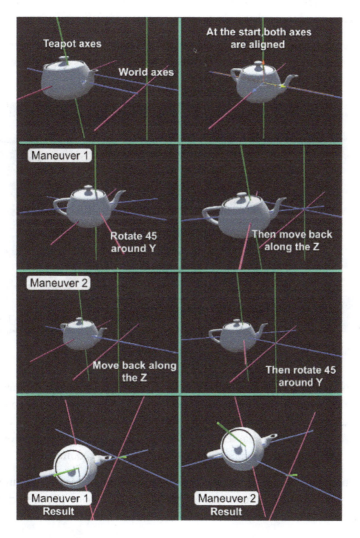

Figure 12.9: Applying transformations in different orders

When an object in OpenGL is manipulated by transformations, it changes its perspective on the world. For an object such as the teapot in *Figure 12.9*, there is an axis system that is its frame of reference, and then there is the world axes system. When a rotation occurs, these become unaligned.

In **Maneuver 1** in *Figure 12.9*, the teapot is rotated 45 degrees around its up axis and then moved back along the *z* axis. After its rotation, the teapot's *z* axis is no longer aligned with the world *z* axis, so when the teapot moves along what it knows to be the *z* axis, it's not moving according to the world *z* axis.

In **Maneuver 2** in *Figure 12.9*, the teapot is moved along the *z* axis first. Its *z* axis and the world's are aligned, and therefore it moves in what you naturally assume is the *z* direction. The rotation happening at the end of the maneuver doesn't change its location.

The result, as can be seen in an overlay of frames for the different maneuvers in *Figure 12.10*, shows how **Maneuver 2** (on the right) will result in the teapot spinning on the spot, whereas **Maneuver 1** (on the left) has its rotation around the origin of the world:

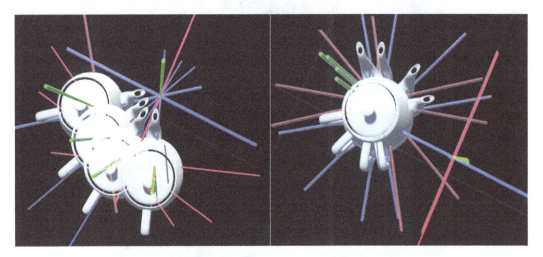

Figure 12.10: A view of the maneuvers with four overlapped frames

What we have explored in this section is how different orders of the same transformation can result in completely different movements of objects. This is something you need to keep in mind when manipulating objects and creating animations. Although we've not explored the order in which scaling is applied, you can imagine that it might, if applied first, shrink the world down such that transformations are applied at a different scale as to what you might require.

One thing to note about how OpenGL processes the transformation lines is that they must be listed in reverse order to what you require. For example, for **Maneuver 1** in *Figure 12.9* to perform a rotation and then a translation, the code listing should be this:

```
glTranslatef(pos.x, pos.y, pos.z)
glRotated(rot_angle, rot_axis.x, rot_axis.y, rot_axis.z)
```

Notice how it is the opposite of what you need. This is again down to the way OpenGL and most other graphics APIs create compound transformations, which we will look at in *Chapter 13, Understanding the Importance of Matrices*.

It's worth noting at this stage that while rotations with Euler angles are straightforward to understand with respect to the angles being rotated around each axis, during some compound rotations, the mathematics breaks down in a seemingly bizarre way. To fully appreciate how this happens, we need to examine transformations and matrices a little closer. Therefore, a discussion on this effect, called **gimbal lock**, will be left until *Chapter 15, Navigating the View Space*.

To complete our discussion of affine transformations before we end the chapter, we will have a look at the final two affine transformations that aren't used a great deal and are not part of the OpenGL API unless programmed manually—shearing and reflections.

Shearing and reflections

The last of the affine transformations are **shearing** and **reflections**. These often don't have a lot of time spent on them in graphics because they aren't often used and aren't part of a typical graphics API. However, for completeness, we will add them here as they generate objects that retain parallelism and ratios.

Shearing is a translation along one dimension. Formally, we can define a point to be sheared thus:

$$P_x = Q_x + sQ_y$$

$$P_y = Q_y$$

This is a two-dimensional case as it can easily be demonstrated in 2D. You will notice that the y coordinate undergoes no change, while the x coordinate is its old self plus a scaled value of the old y coordinate. It can, of course, also be the other way around, with x staying the same and y changing. The result of shearing is to make an object appear to lean over, as demonstrated in *Figure 12.11*. It's a donut shape, shown as the original shape on the left and the sheared version on the right. To shear it, all the vertices are selected and then dragged in just one direction:

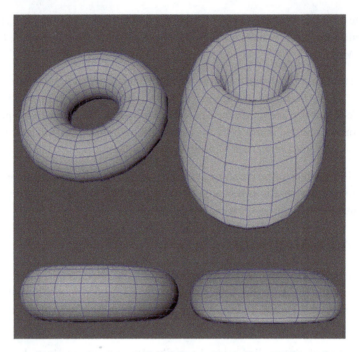

Figure 12.11: A donut that has been sheared in one direction

The other less-used affine transformation is that of reflection. Its result is as you would expect. It takes an object and reverses the coordinates in one direction to create a mirror image. As with shearing, there's no dedicated method in OpenGL to achieve it, though it can be achieved by multiplying one dimension of an object by -1. As shown in *Figure 12.12*, a reflection can be easily achieved in a 3D modeling package by multiplying all coordinates in one dimension by -1:

Figure 12.12: The reflection of a teapot created by multiplying the z coordinates of the vertices by -1

You can, of course, achieve this in OpenGL quite easily, and you've probably already figured out how to do it. Something like this would work:

```
glScaled(1, 1, -1)
```

However, if you do try this, be warned that the object will turn inside out, and you won't be able to see it unless rendering in wireframe mode, and the mesh is textured with backface culling off.

Summary

Along with the data structures required for storing and drawing a mesh, affine transformations are a fundamental part of computer graphics. In fact, how OpenGL stores and manipulates them is the crux of all graphics processing.

In this chapter, we've examined the five transformations that fit into this special category—translation, scaling, rotation, shearing, and reflection. Of all these, rotation is the most complex. The mathematics of rotations has been a hotly researched topic in computer graphics since its inception, as you too will experience as we progress through the book.

We began by reviewing translation, a functionality we have been using since the creation of the `Transform` class in our Python project. It, along with scaling, is a very simple mathematical function. We spent a lot of time adding the final touches to the `Transform` class, completing it with methods to control scaling and rotation.

These newfound skills will assist you in expanding your knowledge of object placement and movement in both 2D and 3D spaces. It's a critical part of your toolkit as a graphics programmer, and you should be aware of how the transformations work and interact with each other. The more you work with them, the more intuitive you will find their use.

Thus far, we've examined the mathematics of these transformations as simple linear equations (although we might not be able to go as far as to call rotational mathematics simple). In this format, they are easy to understand and less abstract than how they are actually stored and represented in graphics using matrices. However, now that you understand the mathematics of affine transformations, it will make the transition to matrices easier. Why do I seem to be attempting to convince you of this? Well, it will become very clear in the next chapter when we'll jump straight into the underlying principle of storage and manipulation mathematics used in every area of graphics.

Answers

Exercise A. (4, 9)

Exercise B. (-1, 5)

Exercise C. (5, 0, 8)

Exercise D.

To `Transform.py`, add the following method:

```
def update_scale(self, amount: pygame.Vector3):
    self.scale.x *= amount.x
    self.scale.y *= amount.y
    self.scale.z *= amount.z
```

Then, call `update_scale()` in the main loop of `ScalingObjects.py`, thus:

```
keys = pygame.key.get_pressed()
if keys[pygame.K_UP]:
    trans.update_scale(pygame.Vector3(2, 2, 2))
if keys[pygame.K_DOWN]:
    trans.update_scale(pygame.Vector3(0.5, 0.5, 0.5))
```

13
Understanding the Importance of Matrices

Matrices are an advanced mathematical concept that you will find everywhere throughout computer games and graphics. Their power comes from their ability to store affine transformations and apply them through multiplication. An understanding of their inherent structure and mathematical operations will provide you with a deep appreciation of the methods underpinning all graphics and game engines.

In this chapter, we will cover the following topics:

- Defining matrices
- Performing operations on matrices
- Creating matrix representations of affine transformations
- Combining transformation matrices for complex maneuvers

We will begin by defining matrices and working through all the mathematical operations that can be performed on them. Then, we will examine how the affine transformations presented in *Chapter 12, Mastering Affine Transformations*, can be achieved through matrix operations. This will involve clarifying the benefits and use of working with matrices in a format called **homogeneous representation**. Finally, you will practice manually calculating transformations with matrices while using your OpenGL Python project to reveal the matrices it uses to manipulate the objects you've been drawing.

By the end of this chapter, you will have discovered how valuable matrices are in computer graphics and be comfortable using them to define compound affine transformations.

Technical requirements

In this chapter, we will be using Python, PyCharm, and Pygame, as used in previous chapters.

Before you begin coding, create a new folder in the PyCharm project for the contents of this chapter called `Chapter_13`.

The solution files for this chapter can be found on GitHub at `https://github.com/ PacktPublishing/Mathematics-for-Game-Programming-and-Computer- Graphics/tree/main/Chapter13`.

Defining matrices

A matrix is an array of numbers. It is defined by the number of rows and columns that define its size. For example, this is a matrix with three rows and two columns:

$$\begin{bmatrix} 3 & 2 \\ 4 & 5 \\ 6 & 1 \end{bmatrix}$$

Each value in the matrix is associated with its location. The value of 3 in the preceding matrix is located in row 0, column 0. More formally, we write the following:

$$M = \begin{bmatrix} 3 & 2 \\ 4 & 5 \\ 6 & 1 \end{bmatrix}$$

$$M[0,0] = 3$$

The values specified in the square brackets are in the order [row, column], like so:

$$M[2,1] = 1$$

$$M[1,0] = 4$$

Here, the value in row 2, column 1 is 1, and the value in row 1, column 0 is 4.

In pure theoretical mathematics, the row and column values start at 1. We are starting our count at 0 because, in programming, when storing arrays and matrices, the index values start at 0.

Now, let's take a look at the mathematical operations that can be achieved with two matrices.

Performing operations on matrices

All matrix manipulation occurs via a set of mathematical operations based on addition, subtraction, and multiplication. Because matrix operations use these familiar fundamentals of arithmetic, they are a relatively easy concept to grasp. In this section, you will explore each of these, beginning with addition, subtraction, and multiplication. However, when it comes to division, as you will soon experience, a whole new set of concepts are required. We will cover these toward the end of this section.

Go easy on yourself if you haven't worked with matrices before and become overwhelmed with the content. They take a lot of practice to become comfortable with and you may not appreciate many of them until you get to apply them in your own graphics projects.

Let's start with the very familiar and simple addition and subtraction operations.

Adding and subtracting matrices

To add or subtract two matrices, they *must* be the same size. Matrices are considered the same size when they have the same number of rows and columns. To add the values, the numbers in the same locations are added together and placed in a resulting matrix of the same size; for example:

$$\begin{bmatrix} 3 & 2 \\ 4 & 5 \\ 6 & 1 \end{bmatrix} + \begin{bmatrix} 1 & 1 \\ 3 & 0 \\ 7 & 3 \end{bmatrix} = \begin{bmatrix} 4 & 3 \\ 7 & 5 \\ 13 & 4 \end{bmatrix}$$

Here, you can see how the value 3 in the first matrix at position [0, 0] is added to the value 1 in [0, 0] in the second matrix, and that the result of 3 + 1 is placed into the solution at [0, 0]. The same operation occurs for all values in both matrices until the solution matrix is complete.

Subtracting matrices occurs in the same way, except that the values in the second matrix are subtracted from those in the first matrix; for example:

$$\begin{bmatrix} 3 & 2 \\ 4 & 5 \\ 6 & 1 \end{bmatrix} - \begin{bmatrix} 1 & 1 \\ 3 & 0 \\ 7 & 3 \end{bmatrix} = \begin{bmatrix} 2 & 1 \\ 1 & 5 \\ -1 & -2 \end{bmatrix}$$

As well as being able to add and subtract matrices, they can be multiplied by a value.

Multiplying by a single value

A matrix can be multiplied by a single number called a *scalar*. This scalar value is multiplied by all the values in the matrix. The size of the resulting matrix is the same as the original one; for example:

$$3 \times \begin{bmatrix} 3 & 2 \\ 4 & 5 \\ 6 & 1 \end{bmatrix} = \begin{bmatrix} 9 & 6 \\ 12 & 15 \\ 18 & 3 \end{bmatrix}$$

Notice that the result goes into the same position in the resulting matrix as the multiplied value from the original. Here, you can see that the value of 2 at position [0, 1], when multiplied by 3 (the scalar), goes into the position where 6 is in the result at [0, 1].

Multiplying by a single value is straightforward. Something a little more complex is multiplying one matrix with another.

Multiplying one matrix with another

Two matrices can only be multiplied together when the first matrix's number of columns is the same as the second matrix's number of rows. In the case of the following example, the first matrix has three columns, and the second matrix has three rows:

$$\begin{bmatrix} 1 & 4 & 3 \\ 2 & 0 & 3 \end{bmatrix} \cdot \begin{bmatrix} 3 & 2 \\ 4 & 5 \\ 6 & 1 \end{bmatrix} = \begin{bmatrix} 37 & 25 \\ 24 & 7 \end{bmatrix}$$

The resulting matrix will have the same number of rows as the first matrix and the same number of columns as the second matrix. In this case, multiplying the first matrix, which has a size of 2 x 3, with the second matrix, which has a size of 3 x 2, will produce the resulting matrix, which is 2 x 2.

Also, note that in the equation, the two matrices do not require a multiplication sign between them. Instead, a single dot is used – not a full stop, but a vertically centered dot. The reason for this is that the operation of multiplying matrices like this is called the dot product. This is the very same operation we used to calculate the angle between vectors in *Chapter 9, Practicing Vector Essentials*. But what does this look like for matrix multiplication? Let's take a look. First, we will strip the matrices of values and use position holders instead, like this:

$$\begin{bmatrix} m_{00} & m_{01} & m_{02} \\ m_{10} & m_{11} & m_{12} \end{bmatrix} \cdot \begin{bmatrix} n_{00} & n_{10} \\ n_{10} & n_{11} \\ n_{20} & n_{21} \end{bmatrix} = \begin{bmatrix} p_{00} & p_{01} \\ p_{10} & p_{11} \end{bmatrix}$$

The calculations take place by working out the dot product between each row in the first matrix with each column in the second matrix, like so:

$$p_{00} = m_{00} \times n_{00} + m_{01} \times n_{10} + m_{02} \times n_{20}$$

$$p_{01} = m_{00} \times n_{10} + m_{01} \times n_{11} + m_{02} \times n_{21}$$

$$p_{10} = m_{10} \times n_{00} + m_{11} \times n_{10} + m_{12} \times n_{20}$$

$$p_{11} = m_{10} \times n_{10} + m_{11} \times n_{11} + m_{12} \times n_{21}$$

In the next section, we will be moving on to something far more complicated to do with matrix operations, so it's probably a good time to pause and reflect on what you've learned so far with a few exercises.

Your turn...

Exercise A. Calculate the following:

$$\begin{bmatrix} 2 & 5 \\ 7 & 6 \end{bmatrix} + \begin{bmatrix} 1 & 9 \\ 8 & 3 \end{bmatrix}$$

Exercise B. Calculate the following:

$$\begin{bmatrix} 2 & 5 \\ -8 & 6 \end{bmatrix} - \begin{bmatrix} -1 & 2 \\ 2 & 3 \end{bmatrix}$$

Exercise C. Calculate the following:

$$4 \times \begin{bmatrix} 3 & 2 \\ 4 & 5 \\ 6 & 1 \end{bmatrix}$$

Exercise D. Calculate the following:

$$\begin{bmatrix} 7 & 2 & 0 \\ 2 & 1 & 0 \end{bmatrix} \cdot \begin{bmatrix} 0 & 2 \\ 4 & 1 \\ 3 & 1 \end{bmatrix}$$

Exercise E. Calculate the following:

$$\begin{bmatrix} 1 & 2 & 3 \\ 4 & 5 & 6 \\ 7 & 8 & 9 \end{bmatrix} \cdot \begin{bmatrix} 9 & 8 & 7 \\ 6 & 5 & 4 \\ 3 & 2 & 1 \end{bmatrix}$$

Dividing matrices

You cannot divide one matrix by another. Truly! There's no equation for performing a division between matrices. Instead, we must perform a multiplication, like this:

$$\frac{M}{N} = M \cdot N^{-1}$$

But what is N^{-1}? It means the **inverse** of the matrix N. If a matrix is multiplied by its inverse, then the resulting matrix is an **identity matrix**, like so:

$$N \cdot N^{-1} = I$$

I bet you have more questions right now. And rightly so. This section is a trip down the rabbit hole in which you are going to learn a lot of new terminology and concepts before we answer the initial question of solving matrix division.

An identity matrix is a **square matrix** with 1s on the diagonal and 0s everywhere else. A square matrix is one where the number of rows equals the number of columns. Therefore, a matrix of size 3 x 3 is considered square. An identity matrix of this size would look like this:

$$\begin{bmatrix} 1 & 0 & 0 \\ 0 & 1 & 0 \\ 0 & 0 & 1 \end{bmatrix}$$

If you're thinking you've heard the word *identity* before in this book, that's because you have – specifically, in *Chapter 7, Interactions with the Keyboard and Mouse for Dynamic Graphics Programs.* Remember this line of code?

```
glLoadIdentity()
```

In OpenGL, it does exactly what its name suggests – it loads an identity matrix onto the **matrix stack**. This is another term that may be new to you, but we won't go into that right now as it will complicate things; we'll leave this until *Chapter 14, Working with Coordinate Spaces.*

The fact that a matrix multiplied by its inverse results in an identity matrix is an interesting fact but it doesn't help us calculate the inverse matrix. Instead, to find the inverse matrix, we have to calculate the matrix's **determinant**.

Calculating the determinant

The determinant is a special value that can be calculated from a square matrix. It is useful in working with linear equations, calculus, and, more importantly for us, finding the inverse of a matrix. For a 2 x 2 matrix, the determinant can be found by multiplying the opposite values, and then taking one result away from the other, like this:

$$det \begin{bmatrix} a & b \\ c & d \end{bmatrix} = a \times d - b \times c$$

The following is an example:

$$det \begin{bmatrix} 3 & 1 \\ 5 & 2 \end{bmatrix} = 3 \times 2 - 1 \times 5 = 6 - 5 = 1$$

If you consider the operation we've just performed visually, we are multiplying values in a criss-cross manner, like this:

$$\begin{bmatrix} a & b \\ c & d \end{bmatrix}$$

Now, there is a sane reason for looking at the operation like this: it will help us work out the determinant for larger matrices as their calculations become more complex. Let's take a look at a 3 x 3 matrix:

$$det \begin{bmatrix} a & b & c \\ d & e & f \\ g & h & i \end{bmatrix} = a(e \times i - f \times h) - b(d \times i - f \times g) + c(d \times h - e \times g)$$

If the calculation looks rather nasty, then consider it visually, as we did with the 2 x 2 matrix:

$$\begin{bmatrix} a & b & c \\ d & e & f \\ g & h & i \end{bmatrix} - \begin{bmatrix} a & b & c \\ d & e & f \\ g & h & i \end{bmatrix} + \begin{bmatrix} a & b & c \\ d & e & f \\ g & h & i \end{bmatrix}$$

You might read this as the determinant using each of the values in the top rows as scalars for the determinants of the 2 x 2 matrix composed of the values in the other rows, which aren't in the same column as the scalar value.

For determinants of larger matrices, check out https://mathinsight.org/determinant_matrix. For a quick determinant calculation, try https://matrix.reshish.com/determinant.php.

Your turn...

Exercise F. Calculate the following:

$$det \begin{bmatrix} 5 & 1 \\ 8 & 3 \end{bmatrix}$$

Exercise G. Calculate the following:

$$det \begin{bmatrix} 3 & 2 & 1 \\ 4 & 1 & 2 \\ 3 & 7 & 6 \end{bmatrix}$$

Calculating the inverse

Now that we can calculate the determinant, how do we use it to find the inverse of a matrix? The formula is as follows:

$$\begin{bmatrix} a & b \\ c & d \end{bmatrix}^{-1} = \frac{1}{ad - bc}\begin{bmatrix} d & -b \\ -c & a \end{bmatrix}$$

Putting this into words, the inverse of a matrix is 1 over the determinant multiplied by a matrix constructed from the original, which has the a and d values swapped and the b and c values negated.

Let's try working with this example:

$$\begin{bmatrix} 1 & 3 \\ 2 & 4 \end{bmatrix}^{-1} = \frac{1}{4-6}\begin{bmatrix} 4 & -3 \\ -2 & 1 \end{bmatrix} = \frac{1}{-2}\begin{bmatrix} 4 & -3 \\ -2 & 1 \end{bmatrix} = -0.5\begin{bmatrix} 4 & -3 \\ -2 & 1 \end{bmatrix} = \begin{bmatrix} -2 & 1.5 \\ 1 & -0.5 \end{bmatrix}$$

How do we know whether this is correct? Well, remember that a matrix multiplied by its inverse will result in the identity matrix. Therefore, we can perform the following calculation to check:

$$\begin{bmatrix} 1 & 3 \\ 2 & 4 \end{bmatrix}\begin{bmatrix} -2 & 1.5 \\ 1 & -0.5 \end{bmatrix} = \begin{bmatrix} 1 \times -2 + 3 \times 1 & 1 \times 1.5 + 3 \times -0.5 \\ 2 \times -2 + 4 \times 1 & 2 \times 1.5 + 4 \times -0.5 \end{bmatrix} = \begin{bmatrix} 1 & 0 \\ 0 & 1 \end{bmatrix}$$

The result will always give you an identity matrix.

Your turn...

Exercise H. Calculate the following:

$$\begin{bmatrix} 1 & 3 \\ 2 & 4 \end{bmatrix}^{-1}$$

Calculating the division

At the beginning of this section, we set out to calculate the division of two matrices, which we later found out wasn't possible. Instead, we need to perform multiplication, like so:

$$\frac{M}{N} = M \cdot N^{-1}$$

Now that we know how to find the inverse of a matrix, we can perform division; for example:

$$\frac{\begin{bmatrix} 4 & 2 \\ 1 & 3 \end{bmatrix}}{\begin{bmatrix} 2 & 5 \\ 5 & 5 \end{bmatrix}} = \begin{bmatrix} 4 & 2 \\ 1 & 3 \end{bmatrix} \cdot \begin{bmatrix} 2 & 5 \\ 5 & 5 \end{bmatrix}^{-1}$$

$$\begin{bmatrix} 2 & 5 \\ 5 & 5 \end{bmatrix}^{-1} = \frac{1}{2 \times 5 - 5 \times 5}\begin{bmatrix} 5 & -5 \\ -5 & 2 \end{bmatrix} = \begin{bmatrix} -0.33 & 0.33 \\ 0.33 & -0.13 \end{bmatrix}$$

$$\begin{bmatrix} 4 & 2 \\ 1 & 3 \end{bmatrix} \cdot \begin{bmatrix} -0.33 & 0.33 \\ 0.33 & -0.13 \end{bmatrix} = \begin{bmatrix} -0.66 & 1.06 \\ 0.66 & -0.06 \end{bmatrix}$$

Instead of explicitly performing a division, the result is obtained by multiplying one matrix with the inverse of the divisor matrix.

Your turn...

Exercise I. Calculate the following:

$$\frac{\begin{bmatrix} 1 & 8 \\ 2 & 3 \end{bmatrix}}{\begin{bmatrix} 5 & 9 \\ 7 & 0 \end{bmatrix}}$$

This section has presented a brief overview of the mathematical operations that can be performed with matrices, as well as revealing the new concepts of determinants and identity matrices. While only about half of this chapter is devoted to examining these concepts, this should give you a powerful skill set that will greatly enhance your ability to work with graphical applications, from animations and shader coding to artificial intelligence. This content is usually delivered to students in several university-level subjects, but unfortunately, there's only limited space in a book that covers mathematics across the breadth of computer graphics to examine them in detail. Having said that, I strongly encourage you to explore the topic further. To this end, I have provided some extra learning resources.

> **Extra learning resources**
>
> Use the following links to practice and strengthen your understanding of matrix mathematics:
>
> `https://www.mathsisfun.com/algebra/matrix-introduction.html`
>
> `https://www.khanacademy.org/math/algebra-home/alg-matrices`
>
> `https://www.cs.mcgill.ca/~rwest/wikispeedia/wpcd/wp/m/Matrix_%2528mathematics%2529.htm`

Thus far, we've examined matrices from a theoretical viewpoint, and you've been able to explore the operations that are used to manipulate them and calculate values. But what do any of the values mean? Without context, they are just fancy data structures.

The true power of matrices cannot be fully appreciated until they are used in a practical setting. They are used throughout computer graphics because of the way they store information, as well as the power that can be achieved when they're multiplied. There's no better place to see this applied than when they are used to represent affine transformations.

Creating matrix representations of affine transformations

In *Chapter 12, Mastering Affine Transformations*, we examined numerous techniques for repositioning and resizing vertices and meshes. The mathematics involved, except for rotations, was mostly straightforward. For these formulae, we applied straightforward arithmetic and some trigonometry to build up equations. Would it surprise you to know that you can represent these transformations as matrix operations? In this section, I will reveal how this can be achieved.

Moving from linear equations to matrix operations

Let's remind ourselves of the formulae used for the most popular of the affine transformations – translation, scaling, and rotation. The point, Q, can be translated by adding a translation value, T, to each of its coordinates, resulting in a new point, P:

$P(x, y, z) = T(x, y, z) + Q(x, y, z)$

We can turn this into a matrix addition operation like so:

$$\begin{bmatrix} P_x \\ P_y \\ P_z \end{bmatrix} = \begin{bmatrix} T_x \\ T_y \\ T_z \end{bmatrix} + \begin{bmatrix} Q_x \\ Q_y \\ Q_z \end{bmatrix}$$

If you are thinking that I've only turned these into arrays and not matrices, then you'd be incorrect. An array is a one-dimensional matrix. It either has one row or one column, depending on its orientation. If you remember back to earlier in this chapter, in the *Adding and subtracting matrices* section, when we looked at additions with matrices, you learned how the values add together to result in the P matrix. In the same way, we performed a translation with linear equations:

$$P = \begin{bmatrix} T_x + Q_x \\ T_y + Q_y \\ T_z + Q_z \end{bmatrix}$$

We can also turn a scaling transformation into a matrix operation. Recall that the scaling formula is as follows:

$P(x, y, z) = S \times Q(x, y, z)$

It can also be written like so:

$P(x, y, z) = S(x, y, z) \times Q(x, y, z)$

This demonstrates that the scale that's applied to x, y, and z of the Q point can be all the same value or different values. This can be turned into a matrix multiplication operation like so:

$$\begin{bmatrix} P_x \\ P_y \\ P_z \end{bmatrix} = \begin{bmatrix} S_x & 0 & 0 \\ 0 & S_y & 0 \\ 0 & 0 & S_z \end{bmatrix} \cdot \begin{bmatrix} Q_x \\ Q_y \\ Q_z \end{bmatrix}$$

Let's expand this using the matrix multiplication method so that you can see the result is to multiply the correct values together to gain a scaling operation:

$$P = \begin{bmatrix} S_x \times Q_x + 0 \times Q_y + 0 \times Q_z \\ 0 \times Q_x + S_y \times Q_y + 0 \times Q_z \\ 0 \times Q_x + 0 \times Q_y + S_z \times Q_z \end{bmatrix} = \begin{bmatrix} S_x \times Q_x \\ S_y \times Q_y \\ S_z \times Q_z \end{bmatrix}$$

So, it's a little more complex than addition, but it does perform scaling with matrices.

Finally, we have rotation. This is inherently more difficult because rotation has three different operations, depending on the axis of rotation. Recall the rotation operations for an X-axis roll:

$$P_x = Q_x$$

$$P_y = Q_y \times \cos(\theta) + Q_z \times \sin(\theta)$$

$$P_z = Q_y \times -\sin(\theta) + Q_z \times \cos(\theta)$$

Just like for scaling, we want to consider the operations that are being performed on each of the x, y, and z coordinates so that we can place the operations in the correct location in a matrix. These operations, when converted into matrix multiplication, turn three separate formulas into one:

$$\begin{bmatrix} P_x \\ P_y \\ P_z \end{bmatrix} = \begin{bmatrix} 1 & 0 & 0 \\ 0 & \cos(\theta) & \sin(\theta) \\ 0 & -\sin(\theta) & \cos(\theta) \end{bmatrix} \cdot \begin{bmatrix} Q_x \\ Q_y \\ Q_z \end{bmatrix}$$

Can you see how the three separate equations for an x pitch are contained within this matrix multiplication? If you do the multiplication, you will end up with this:

$$P = \begin{bmatrix} Q_x \\ Q_y \times \cos(\theta) + Q_z \times \sin(\theta) \\ Q_y \times -\sin(\theta) + Q_z \times \cos(\theta) \end{bmatrix}$$

The separate formulas for the Y-axis are as follows:

$$P_x = Q_x \times \cos(\theta) - Q_z \times \sin(\theta)$$

$$P_y = Q_y$$

$$P_z = Q_x \times \sin(\theta) + Q_z \times \cos(\theta)$$

When the Y-axis rotation formulas are condensed into a matrix multiplication, we get the following:

$$\begin{bmatrix} P_x \\ P_y \\ P_z \end{bmatrix} = \begin{bmatrix} \cos(\theta) & 0 & -\sin(\theta) \\ 0 & 1 & 0 \\ \sin(\theta) & 0 & \cos(\theta) \end{bmatrix} \cdot \begin{bmatrix} Q_x \\ Q_y \\ Q_z \end{bmatrix}$$

Finally, we can take the equations for a Z-axis roll:

$$P_x = Q_x \times \cos(\theta) + Q_y \times \sin(\theta)$$

$$P_y = Q_x \times -\sin(\theta) + Q_y \times \cos(\theta)$$

$$P_z = Q_z$$

When the Z-axis rotation formulae are merged into a matrix multiplication, we get the following:

$$\begin{bmatrix} P_x \\ P_y \\ P_z \end{bmatrix} = \begin{bmatrix} \cos(\theta) & \sin(\theta) & 0 \\ -\sin(\theta) & \cos(\theta) & 0 \\ 0 & 0 & 1 \end{bmatrix} \cdot \begin{bmatrix} Q_x \\ Q_y \\ Q_z \end{bmatrix}$$

Wow – a lot of mathematics has just been presented in this section. Take your time understanding it as it's an exceptionally important concept in computer graphics. It might seem like a lot of fuss over nothing and a way to complicate all the calculations, but it endeavours to make them easier, as you will see in the next section.

Compounding affine transformations

Thus far, we have converted the affine transformation equations that were revealed in *Chapter 12, Mastering Affine Transformations*, into matrix operations. Take another look at them. What is similar or not similar in the results for the individual transformations? The difference I'd like you to spot is that the translate matrix operations involve adding one-dimensional matrices, whereas the scaling and rotation operations are multiplications with 3 x 3 matrices.

If you didn't know, the fastest mathematical operation a computer can perform is multiplication. So, if we want to start mixing translation, scaling, and rotation, it would make sense for the three to be in the same format so that we can achieve smooth multiplication between them. In OpenGL, when a series of glTranslate(), glScale(), and glRotate() commands are executed, they are multiplied together into the same matrix. So, how does this happen?

Enter **homogeneous representation**. By adding a fourth component to the end of points and vectors to create what's called a **homogenous coordinate**, all three transformation matrices can be multiplied. This works by adding a value of 0 to the end of a vector's representation and a 1 to a point. For example, the vector (8, 3, 4) becomes (8, 3, 4, 0) and the point (1, 4, 3) becomes (1, 4, 3, 1). The last component is given a designation of w so that a vector or point is represented by (x, y, z, w).

How does this change the transformation matrices? Well, the translation becomes as follows:

$$\begin{bmatrix} P_x \\ P_y \\ P_z \\ P_w \end{bmatrix} = \begin{bmatrix} 1 & 0 & 0 & T_x \\ 0 & 1 & 0 & T_y \\ 0 & 0 & 1 & T_z \\ 0 & 0 & 0 & 1 \end{bmatrix} \cdot \begin{bmatrix} Q_x \\ Q_y \\ Q_z \\ Q_w \end{bmatrix}$$

Expanding this, you can see how the addition operation of translation is maintained:

$$P = \begin{bmatrix} 1 \times Q_x + 0 \times Q_y + 0 \times Q_z + T_x \times Q_w \\ 0 \times Q_x + 1 \times Q_y + 0 \times Q_z + T_y \times Q_w \\ 0 \times Q_x + 0 \times Q_y + 1 \times Q_z + T_z \times Q_w \\ 0 \times Q_x + 0 \times Q_y + 0 \times Q_z + 1 \times Q_w \end{bmatrix} = \begin{bmatrix} Q_x + T_x \times Q_w \\ Q_y + T_y \times Q_w \\ Q_z + T_z \times Q_w \\ Q_w \end{bmatrix}$$

If a point is being translated, the value of Q_w will be 1 and the beforehand matrix can be simplified like so:

$$\begin{bmatrix} Q_x + T_x \\ Q_y + T_y \\ Q_z + T_z \\ 1 \end{bmatrix}$$

If a vector is being translated, the value of Q_w will be 0 and the beforehand matrix can be simplified like so:

$$\begin{bmatrix} Q_x \\ Q_y \\ Q_z \\ 0 \end{bmatrix}$$

As you can see, a point will be moved, whereas a vector will retain its original value. This works with point and vector mathematics as points can be moved, but vectors can't as they don't represent a location in space. If you were to move a vector, it would retain its original value as it simply represents direction and magnitude, not location.

Now that the translation operations have been converted into a homogeneous representation, it can be multiplied by the scaling and rotation matrices in any required order. However, before this can happen, both the scaling and rotation matrices also need to become homogenous. For scaling, the operation becomes as follows:

$$\begin{bmatrix} P_x \\ P_y \\ P_z \\ P_w \end{bmatrix} = \begin{bmatrix} S_x & 0 & 0 & 0 \\ 0 & S_y & 0 & 0 \\ 0 & 0 & S_z & 0 \\ 0 & 0 & 0 & 1 \end{bmatrix} \cdot \begin{bmatrix} Q_x \\ Q_y \\ Q_z \\ Q_w \end{bmatrix}$$

The following is for an X-pitch rotation:

$$\begin{bmatrix} P_x \\ P_y \\ P_z \\ P_w \end{bmatrix} = \begin{bmatrix} 1 & 0 & 0 & 0 \\ 0 & \cos(\theta) & \sin(\theta) & 0 \\ 0 & -\sin(\theta) & \cos(\theta) & 0 \\ 0 & 0 & 0 & 1 \end{bmatrix} \cdot \begin{bmatrix} Q_x \\ Q_y \\ Q_z \\ Q_w \end{bmatrix}$$

For a Y-yaw rotation:

$$\begin{bmatrix} P_x \\ P_y \\ P_z \\ P_w \end{bmatrix} = \begin{bmatrix} \cos(\theta) & 0 & -\sin(\theta) & 0 \\ 0 & 1 & 0 & 0 \\ -\sin(\theta) & 0 & \cos(\theta) & 0 \\ 0 & 0 & 0 & 1 \end{bmatrix} \cdot \begin{bmatrix} Q_x \\ Q_y \\ Q_z \\ Q_w \end{bmatrix}$$

For a Z-roll rotation:

$$\begin{bmatrix} P_x \\ P_y \\ P_z \\ P_w \end{bmatrix} = \begin{bmatrix} \cos(\theta) & \sin(\theta) & 0 & 0 \\ -\sin(\theta) & \cos(\theta) & 0 & 0 \\ 0 & 0 & 1 & 0 \\ 0 & 0 & 0 & 1 \end{bmatrix} \cdot \begin{bmatrix} Q_x \\ Q_y \\ Q_z \\ Q_w \end{bmatrix}$$

Now that you know how to represent the transformation functions in homogeneous coordinates, it's time to examine how they can be multiplied together.

Combining transformation matrices for complex maneuvers

As with the OpenGL order of transformations, which we discussed in *Chapter 12, Mastering Affine Transformations*, when combining these homogeneous representation matrices to produce compound movements involving translation, scaling, and rotation, the matrices are presented in reverse order. For example, to transform a point by (3, 4, 5), rotate it around the X-axis by 45 degrees, and then scale it by 0.3 in all directions; the matrix multiplication is as follows:

$$\begin{bmatrix} 0.3 & 0 & 0 & 0 \\ 0 & 0.3 & 0 & 0 \\ 0 & 0 & 0.3 & 0 \\ 0 & 0 & 0 & 1 \end{bmatrix} \cdot \begin{bmatrix} 1 & 0 & 0 & 0 \\ 0 & \cos(45) & \sin(45) & 0 \\ 0 & -\sin(45) & \cos(45) & 0 \\ 0 & 0 & 0 & 1 \end{bmatrix} \cdot \begin{bmatrix} 1 & 0 & 0 & 3 \\ 0 & 1 & 0 & 4 \\ 0 & 0 & 1 & 5 \\ 0 & 0 & 0 & 1 \end{bmatrix}$$

Note how the translation matrix of the first operation is placed on the right and the scaling matrix on the left. To multiply this out, we begin by multiplying the last two matrices (the translation and rotation) to get the following:

$$
\begin{bmatrix} 0.3 & 0 & 0 & 0 \\ 0 & 0.3 & 0 & 0 \\ 0 & 0 & 0.3 & 0 \\ 0 & 0 & 0 & 1 \end{bmatrix} \cdot \begin{bmatrix} 1 & 0 & 0 & 3 \\ 0 & 0.707 & 0.707 & 6.363 \\ 0 & -0.707 & 0.707 & 0.707 \\ 0 & 0 & 0 & 1 \end{bmatrix}
$$

Then, we complete the multiplication with the remaining two matrices, which results in the following:

$$
\begin{bmatrix} 0.3 & 0 & 0 & 0.9 \\ 0 & 0.2121 & 0.2121 & 1.91 \\ 0 & -0.2121 & 0.2121 & 0.2121 \\ 0 & 0 & 0 & 1 \end{bmatrix}
$$

Although learning to calculate these operations by hand is a great skill to have and will help embed your understanding of them, I'm not going to ask you to do it here. You will, however, now get a chance to examine these operations in OpenGL and use an online calculator to validate the results.

Let's do it...

In this practical exercise, you will create a transformation in your OpenGL Python project and compare the values obtained in the matrices in the program with those calculated manually:

1. Create a new Python folder called `Chapter_13` and copy the contents from `Chapter_12` into it.

2. Make a copy of `ExploreNormals.py` and call it `TransformationMatrices.py`.

3. Modify the code in `TransformationMatrices.py` to draw just a single textured cube positioned at the origin:

```
import math

from Object import *
from pygame.locals import *
from OpenGL.GLU import *
from Cube import *
```

In the first part of the code, notice the reduction in the number of required libraries.

In the second part of the code, ensure you have only one cube being drawn, as follows:

```
. .
done = False
white = pygame.Color(255, 255, 255)

objects_3d = []
```

```
objects_2d = []

cube = Object("Cube")
cube.add_component(Transform((0, 0, 0)))
cube.add_component(Cube(GL_POLYGON,
                        "images/wall.tif"))

objects_3d.append(cube)

clock = pygame.time.Clock()
fps = 30
..
```

Ensure you position it at (0,0,0), and use whatever image you've been putting on the cubes thus far, instead of the wall.tif one that I am using.

The rest of the program remains the same.

When the code is run, the window will be filled with the texture of the cube when the camera is near it, as shown in *Figure 13.1*:

Figure 13.1: Render of a cube at the origin

4. In `Object.py`, re-instantiate the scaling, rotation, and translation lines if you commented them out previously:

```python
def update(self, events = None):
    glPushMatrix()
    for c in self.components:
        if isinstance(c, Transform):
            pos = c.get_position()
            scale = c.get_scale()
            rot_angle = c.get_rotation_angle()
            rot_axis = c.get_rotation_axis()

            glTranslatef(pos.x, pos.y, pos.z)
            glRotated(rot_angle, rot_axis.x,
                        rot_axis.y, rot_axis.z)
            glScalef(scale.x, scale.y, scale.z)

        elif isinstance(c, Mesh3D):
            glColor(1, 1, 1)
```

5. Back in `TransformationMatrix.py`, add the following transformations to the cube:

```python
..
cube = Object("Cube")
cube.add_component(Transform((0, 0, 0)))
cube.add_component(Cube(GL_POLYGON,
                    "images/wall.tif"))
trans: Transform = cube.get_component(Transform)
trans.set_position((0, 0, -3))
trans.set_rotation_axis(pygame.Vector3(1, 0, 0))
trans.update_rotation_angle(45)
trans.set_scale(pygame.Vector3(0.5, 2, 1))

objects_3d.append(cube)

clock = pygame.time.Clock()
..
```

When you run the code now, the cube will have moved slightly back into the screen, rotated by 45 degrees around the *X*-axis, halved in the *x* direction, and doubled the size in the *y* direction, as shown in *Figure 13.2*:

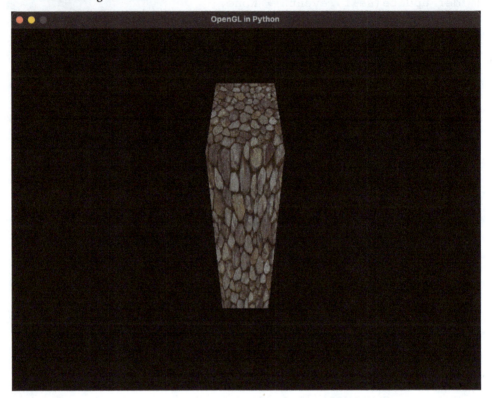

Figure 13.2: The cube after adding transformations

6. Take a look at Object.py and look at the order of execution of the translation, rotation, and scaling:

```
glTranslatef(pos.x, pos.y, pos.z)
glRotated(rot_angle, rot_axis.x,
          rot_axis.y, rot_axis.z)
glScalef(scale.x, scale.y, scale.z)
```

Given that these will execute in reverse order, the matrix multiplication with the values you fed into these operations in *step 5* will be as follows:

$$\begin{bmatrix} 1 & 0 & 0 & 0 \\ 0 & 1 & 0 & 0 \\ 0 & 0 & 1 & -3 \\ 0 & 0 & 0 & 1 \end{bmatrix} \cdot \begin{bmatrix} 1 & 0 & 0 & 0 \\ 0 & \cos(45) & \sin(45) & 0 \\ 0 & -\sin(45) & \cos(45) & 0 \\ 0 & 0 & 0 & 1 \end{bmatrix} \cdot \begin{bmatrix} 0.5 & 0 & 0 & 0 \\ 0 & 2 & 0 & 0 \\ 0 & 0 & 1 & 0 \\ 0 & 0 & 0 & 1 \end{bmatrix}$$

7. You can calculate the result of the multiplication given in *step 6* manually or use the handy online *Matrix Multiplication Calculator* tool available at https://matrix.reshish.com/multCalculation.php.

Because there are three matrices to multiply, we begin with the two on the right. Although I've said matrix multiplication happens backward, here, we start with the two right-most matrices and multiply these going from left to right, as shown in *Figure 13.3*:

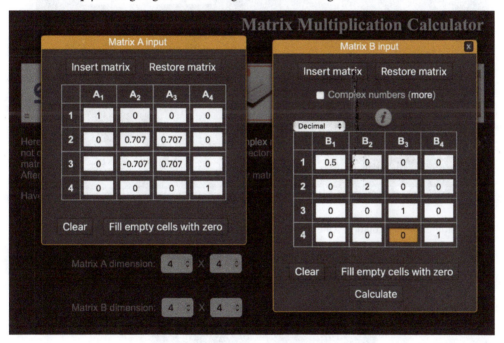

Figure 13.3: Calculating rotation with translation

For this first calculation, note that both matrices are 4 x 4 in size and the cosine and sine, which are at 45 degrees, have already been determined as 0.707. Once this operation is complete, the resulting matrix can be multiplied with the scaling matrix, as shown in *Figure 13.4*. Ensure that you put the scaling matrix to the left:

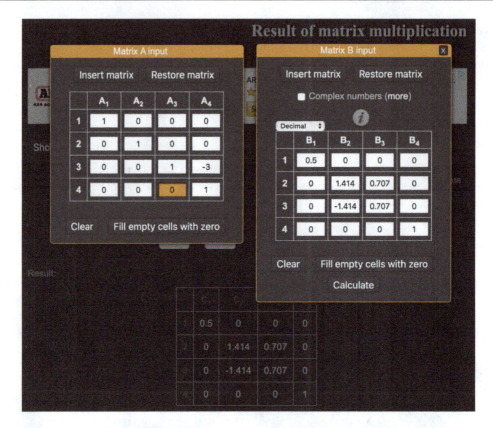

Figure 13.4: Calculating scale with the pre-calculated rotation/translation matrix

On calculating this, the matrix shown in *Figure 13.5* will be displayed:

	C_1	C_2	C_3	C_4
1	0.5	0	0	0
2	0	1.414	0.707	0
3	0	-1.414	0.707	-3
4	0	0	0	1

Figure 13.5: The result of multiplying the translation, rotation, and scaling matrices

8. To validate this calculation, we can also ask OpenGL what it has stored in our program. The internal OpenGL matrix that holds all the transformation multiplications is called the **ModelView Matrix**. We can obtain its value by adding the following code to `Object.py` after the transformations have been applied:

```
rot_axis = c.get_rotation_axis()

glTranslatef(pos.x, pos.y, pos.z)
glRotated(rot_angle, rot_axis.x, rot_axis.y,
          rot_axis.z)
glScalef(scale.x, scale.y, scale.z)
mv = glGetDoublev(GL_MODELVIEW_MATRIX)
print("MV: ")
print(mv)


elif isinstance(c, Mesh3D):
```

After adding this code, run the program. The ModelView Matrix will display on a loop in the console. Once you see it, you can stop running the program. In the console, you should see the following output:

```
MV:
[[ 0.5         0.          0.          0.        ]
 [ 0.          1.41421354  1.41421354  0.        ]
 [ 0.         -0.70710677  0.70710677  0.        ]
 [ 0.          0.         -3.          1.        ]]
```

The matrix will be transposed (the rows and columns will be switched) as that's how OpenGL works with them, but you'll see the results are identical to those we calculated by hand.

As you've seen in this section, matrices are a powerful concept for storing transformational information for 3D objects. While they are laborious to calculate manually, it can be a valuable exercise to sometimes question the outputs from a program against hand-performed calculations and vice versa, to assist you in catching any potential errors. That was the purpose of the practical exercise you've just completed.

Summary

A lot of mathematical concepts were covered in this chapter that focused on matrices. Besides understanding vectors, a solid knowledge of matrices (especially 4 x 4) is an essential skill to have as a graphics programmer since they underpin the majority of the mathematics found in graphics and game engines. Once you appreciate the beauty of their simplicity and power, you'll become more and more comfortable with their use.

In this chapter, we have only scratched the surface of using matrices in graphics. After learning how the addition operation that's used in translations can be transformed into a 4 x 4 matrix, and integrated with scaling and rotation to perform compound transformations in 3D, we took a brief look at the ModelView Matrix in OpenGL using the project code created thus far. However, the way we currently perform the transformations is restricted to the same order as how `glTranslate()`, `glRotate()`, and `glScale()` are used in the existing code.

In the next chapter, we will dig deeper into the matrices used in OpenGL for manipulating objects – not only with the transformations we've been working with thus far but also the camera position and orientation, the projection modes, and allowing more complex compound transformations not restricted by a set coding order. These are the next essential steps in your learning journey, as you'll be moving toward understanding the coordinate spaces used in graphics for displaying 3D objects. This will give you an appreciation of advanced vertex shaders.

Answers

For a great online matrix calculator that will also reveal the working out for you, visit `https://matrix.reshish.com/multiplication.php`:

Exercise A:

$$\begin{bmatrix} 3 & 14 \\ 15 & 9 \end{bmatrix}$$

Exercise B:

$$\begin{bmatrix} 3 & 3 \\ -10 & -11 \end{bmatrix}$$

Exercise C:

$$\begin{bmatrix} 12 & 8 \\ 16 & 20 \\ 24 & 4 \end{bmatrix}$$

Exercise D:

$$\begin{bmatrix} 8 & 16 \\ 4 & 5 \end{bmatrix}$$

Exercise E:

$$\begin{bmatrix} 30 & 24 & 18 \\ 84 & 69 & 54 \\ 138 & 114 & 90 \end{bmatrix}$$

Exercise F:

$$det \begin{bmatrix} 5 & 1 \\ 8 & 3 \end{bmatrix} = 5 \times 3 - 1 \times 8 = 7$$

Exercise G:

$$det \begin{bmatrix} 3 & 2 & 1 \\ 4 & 1 & 2 \\ 3 & 7 & 6 \end{bmatrix} = 3(1 \times 6 - 2 \times 7) - 2(4 \times 6 - 2 \times 3) + 1(4 \times 7 - 1 \times 3) = -35$$

Exercise H:

$$\begin{bmatrix} 4 & 3 \\ 1 & 2 \end{bmatrix}^{-1} = \frac{1}{4 \times 2 - 3 \times 1} \begin{bmatrix} 4 & -3 \\ -2 & 1 \end{bmatrix} = \frac{1}{5} \begin{bmatrix} 4 & -3 \\ -2 & 1 \end{bmatrix} = 0.2 \begin{bmatrix} 4 & -3 \\ -2 & 1 \end{bmatrix} = \begin{bmatrix} 0.4 & -0.6 \\ -0.2 & 0.8 \end{bmatrix}$$

Exercise I:

$$\frac{\begin{bmatrix} 1 & 8 \\ 2 & 3 \end{bmatrix}}{\begin{bmatrix} 5 & 9 \\ 7 & 0 \end{bmatrix}} = \begin{bmatrix} 1 & 8 \\ 2 & 3 \end{bmatrix} \cdot \begin{bmatrix} 5 & 9 \\ 7 & 0 \end{bmatrix}^{-1}$$

$$\begin{bmatrix} 5 & 9 \\ 7 & 0 \end{bmatrix}^{-1} = \frac{1}{5 \times 0 - 9 \times 7} \begin{bmatrix} 0 & -9 \\ -7 & 5 \end{bmatrix} = \begin{bmatrix} 0 & 0.14 \\ 0.11 & -0.08 \end{bmatrix}$$

$$\begin{bmatrix} 1 & 8 \\ 2 & 3 \end{bmatrix} \cdot \begin{bmatrix} 0 & 0.14 \\ 0.11 & -0.08 \end{bmatrix} = \begin{bmatrix} 0.88 & -0.5 \\ 0.33 & 0.04 \end{bmatrix}$$

14

Working with Coordinate Spaces

Understanding the different coordinate spaces used in graphics rendering is a critical and transferable skill that you as a programmer require. These are a universal concept across all graphics and game engines, and being able to apply them to manipulate a virtual scene is a skill you'll never regret acquiring.

These key matrices form the OpenGL matrix stack that defines all mathematical operations. Mathematical operations are required to take the vertices of a model from their own local coordinate system into a pixel on the computer screen. They define not only where individual objects are in a scene and how they are scaled and rotated, but also allow for the creation of a virtual camera. This camera can be moved and orientated to influence the location and orientation from which a scene is viewed.

The modelview, view, and projection matrices contain indispensable mathematical functions for any graphics engine. You've been exploring these mathematical concepts throughout this book. Think of this chapter as a culmination of all the mathematical content you've explored thus far.

In this chapter, we will cover the following topics:

- Understanding OpenGL's Matrix Stack
- Working with the Modelview Matrix
- Working with the View Matrix
- Working with the Projection Matrix

By the end of this chapter, you will have developed the skills to create, manipulate, and apply the matrix stack in 2D and 3D spaces as well as validate the results of the mathematics you are using to develop more robust code and graphics applications.

Technical requirements

In this chapter, we will be using Python, PyCharm, and Pygame, as used in previous chapters.

Before you begin coding, create a new `Chapter_14` folder in the PyCharm project for the contents of this chapter.

The solution files containing the code can be found on GitHub at `https://github.com/ PacktPublishing/Mathematics-for-Game-Programming-and-Computer- Graphics/tree/main/Chapter14`.

Understanding OpenGL's Matrix Stack

In graphics, the current transformation applied to a point or object is determined by the current **model-view-projection** (**MVP**) matrix. This is a culmination of the model matrix, the view matrix, and the projection matrix. We first discussed these matrices as coordinate spaces in *Chapter 4, Graphics and Game Engine Components*. Each one has a specific use in the graphics pipeline, as is shown in *Figure 14.1* (this diagram has been reinserted here from *Chapter 4, Graphics and Game Engine Components*, for your convenience):

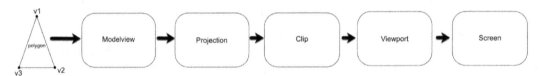

Figure 14.1: The graphics pipeline

The coordinates or points that define a graphics object are stored in a model's local coordinate system. They define the geometry of the object independently of where it is situated in world space. As we saw in *Chapter 4, Graphics and Game Engine Components*, a cube can be defined by six points, one for each of the vertices, thus:

```
cube.vertices = [(0.5, -0.5, 0.5),
                 (-0.5, -0.5, 0.5),
                 (0.5, 0.5, 0.5),
                 (-0.5, 0.5, 0.5),
                 (0.5, 0.5, -0.5),
                 (-0.5, 0.5, -0.5)
                ]
```

These values tell us nothing about where the cube is located in the world, but just the structure of it as it was created. When the cube is placed in the world, transformations are applied that define its location, scale, and orientation. These transformations (accumulated in the modelview) change the values of the vertices so that they can be drawn by the graphics API in the correct location. However, while this process positions the cube in the 3D world (or 2D world if you aren't working with a *z* axis), it doesn't define what the value of each vertex is with respect to the camera viewing it. So, the points undergo another operation that projects them into camera space. Last but not least, the points in camera space need to become pixels on the screen, and therefore they undergo another operation that projects them onto a 2D plane that represents the computer screen.

We could, in fact, say that a rendered cube has four sets of coordinates: one in model space, one in world space, one in eye (or camera) space, and one in screen space. What these coordinates are in each space depends on how the cube sits with respect to the space's origin. This is illustrated in *Figure 14.2*:

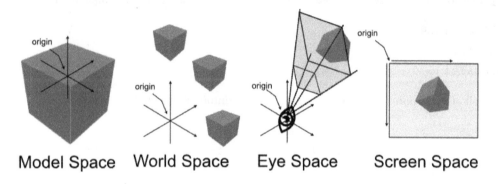

Figure 14.2: Coordinate spaces

When we work in games and graphics, the majority of the operations performed are in world space as it is this virtual world we create and want to position objects within. When you translate, scale, or rotate an object, it is this space you are working in. However, at some point, the vertices of a model end up being converted into each of these spaces to get the final pixels on the screen. Mathematically the entire process that draws a coordinate ,Q, on the screen from an original point, *P*, looks like this:

```
Q = projection matrix * view matrix * model matrix * P
```

The model matrix puts the point into world space, the view matrix then places it in eye space, and then the projection matrix converts it into screen coordinates. Notice how the operation works entirely on multiplication? This is what makes these matrices so efficient, as we discussed in *Chapter 13, Understanding the Importance of Matrices*. We will now look at each of the matrices that comprise the OpenGL matrix stack in turn to gain a greater insight into their purpose.

Working with the Model Matrix

The model matrix is the accumulation of the multiplications of the transformation matrices that are to be applied to a point or vector. As we discovered in *Chapter 13, Understanding the Importance of Matrices*, the order in which the transformations are multiplied is important to the final outcome. Also, at the end of the same chapter, you discovered that in OpenGL, you can obtain the contents of the modelview matrix with the following code:

```
glGetDoublev(GL_MODELVIEW_MATRIX)
```

Rather than using OpenGL's own methods for moving, resizing, and orienting an object, you can set the matrix manually, and then perform matrix multiplication to apply the transformation to a model as long as you keep in mind the format of the modelview matrix.

These transformations were performed in *Chapter 13, Understanding the Importance of Matrices*:

```
glTranslatef(0, 0, -3)
glRotated(45, 1, 0, 0)
glScalef(0.5, 2, 1)
```

The result of performing these transformations was a printout of this modelview matrix:

```
[[ 0.5          0.          0.          0.          ]
 [ 0.          1.41421354  1.41421354  0.          ]
 [ 0.         -0.70710677  0.70710677  0.          ]
 [ 0.          0.         -3.          1.          ]]
```

If you take notice of where the resulting values end up in the matrix, you can see that the

0.   0.   -3. values are the translation values. This is because the 4x4 translation matrix stores its values in the last column (or row, depending on which way around you have the matrix). The 1.41421354   1.41421354 and -0.70710677   0.70710677 values are an accumulation of the rotation and scaling operations, and 0.5 is the *x* scaling factor. When you multiply translations, rotations, and scalings together, you can always determine the values of the translation from the resulting matrix because of their location, whereas the values for the scaling and rotation will always be mixed together. So, it is impossible to tell what the original rotations or scalings were. It's rare that you would want to set the modelview matrix by hand, as it's much easier to calculate it.

Let's give it a go for ourselves.

Let's do it...

In this practical exercise, you will replace the OpenGL methods for translation, scaling, and rotation with your own calculations. Follow these steps:

1. Create a new Python folder called Chapter_14 and copy the contents from Chapter_13 into it.

2. We are going to rewrite the entire contents of the Transform.py class to work with matrices. Delete all the lines of code in your Transform.py class and add the following code:

```python
import pygame
import math
import numpy as np


class Transform:
    def __init__(self):
        self.MVM = np.identity(4)

    def get_MVM(self):
        return self.MVM
```

To begin, the initialization method has had the original variables removed and self.MVM has been added to store the modelview matrix. It is initially set to an empty identity matrix. When we zero a matrix in computer graphics, it's not set to all values of zero; otherwise, any other matrix multiplied with it will result in another matrix full of zeros.

Following this, a new method for returning the modelview matrix has been added.

3. The next method that you add here will update the position of the object. It does this through a multiplication operation of the existing modelview matrix with a translation matrix:

```python
def update_position(self, position):
    self.MVM = self.MVM @ np.matrix([[1, 0, 0, 0],
        [0, 1, 0, 0],
        [0, 0, 1, 0],
        [position.x, position.y,
        position.z, 1]])
```

The `update_scale()` method is similar to the `update_position()` method in that it performs a matrix calculation. But notice the format of the matrix is different to cater to scaling operations:

```python
def update_scale(self, amount: pygame.Vector3):
    self.MVM = self.MVM @ np.matrix([
        [amount.x, 0, 0, 0],
        [0, amount.y, 0, 0],
        [0, 0, amount.z, 0],
        [0, 0, 0, 1]
    ])
```

4. The final three methods you will add are all for rotation. They allow you to rotate around any axis:

```python
def rotate_x(self, amount):
    amount = math.radians(amount)
    self.MVM = self.MVM @ np.matrix([
            [1, 0, 0, 0],
            [0, math.cos(amount),
             math.sin(amount), 0],
            [0, -math.sin(amount),
             math.cos(amount), 0],
            [0, 0, 0, 1]])

def rotate_y(self, amount):
    amount = math.radians(amount)
    self.MVM = self.MVM @ np.matrix([
            [math.cos(amount), 0,
             -math.sin(amount), 0],
            [0, 1, 0, 0],
            [math.sin(amount), 0,
             math.cos(amount), 0],
            [0, 0, 0, 1]])

def rotate_z(self, amount):
```

```
            amount = math.radians(amount)
            self.MVM = self.MVM @ np.matrix([
                    [math.cos(amount), math.sin(amount),
                     0, 0],
                    [-math.sin(amount), math.cos(amount),
                     0, 0],
                    [0, 0, 1, 0],
                    [0, 0, 0, 1]])
```

Take note of how the three rotation matrices have been formatted for the different axes.

5. To use the modelview matrix to set the transformation of the object, the code in `Object. py` needs to be updated thus:

```
    ..
    def update(self, events = None):
        glPushMatrix()
        for c in self.components:
            if isinstance(c, Transform):
                glLoadMatrixf(c.get_MVM())
                mv = glGetDoublev(GL_MODELVIEW_MATRIX)
                print("MV: ")
                print(mv)

            elif isinstance(c, Mesh3D):
                glColor(1, 1, 1)
    ..
```

Take note here of how the OpenGL transformations have been removed and replaced with a `glLoadMatrixf()` call instead. This method loads in the matrix calculated by the `Transform` class. The printing of the modelview matrix has been left to show you the contents of the matrix after we have calculated it manually. The idea is that if you perform the exact same transformations as we did in *Chapter 13, Understanding the Importance of Matrices*, the modelview matrix will be the same.

6. To use these modifications, open up `TransformationMatrices.py` and modify how the cube is being drawn thus:

```
    ..
    objects_3d = []
    objects_2d = []
```

```
cube = Object("Cube")
cube.add_component(Transform())
cube.add_component(Cube(GL_POLYGON, "images/wall.tif"))
trans: Transform = cube.get_component(Transform)

trans.update_position(pygame.Vector3(0, 0, -3))
trans.rotate_x(45)
trans.update_scale(pygame.Vector3(0.5, 2, 1))

objects_3d.append(cube)

clock = pygame.time.Clock()
..
```

In these changes, the Transform component being added to the cube no longer has any parameters passed to it, and the position, rotation, and scaling code has changed to the new methods we just created in Transform.py. The values being used are identical to the ones that we used previously in *Chapter 13, Understanding the Importance of Matrices.*

At this point, you can run TransformationMatrices.py and take a look at the modelview matrix that prints out in the console. Is it the same as the one that OpenGL created in *Chapter 13, Understanding the Importance of Matrices*? It shouldn't be, as I intentionally had you put this code in to point out something. Note that there won't be anything visible in the window; it will just be black.

Take some time now that you have this code working to ponder my last question before you continue reading.

At the beginning of this chapter, I reminded you that the order of transformations matters. Although you may have the position, rotation, and scaling happening in the same order as the previous OpenGL methods used, now that we are calculating them, we need to reverse the order around.

As OpenGL is given each matrix to apply, it places it on a matrix stack. You can imagine this as a tower being constructed where each matrix is a brick placed on top of the previous one. The last matrix added is the topmost one. To process these matrices, OpenGL multiplies them together, starting at the top of the stack and working down. In fact, OpenGL maintains a matrix stack for both the modelview and projection modes and another one for textures. When a command such as glTranslate() is run, a matrix is pushed onto the stack. Whatever is in the stack at the time of drawing a frame gets applied to the geometry of the objects in the scene. When a glLoadIdentity() call is made, OpenGL is essentially zeroing or reinitializing the stack. When the vertices of an object are processed through the stack, the multiplication equation for all matrices is constructed from the bottom of the stack upward. For example, if we perform a translation, a rotation, and then a scale using OpenGL methods, the translation goes on the stack first, followed by the rotation, and then the scale on the top. Reading this in reverse order to construct the multiplication gives the following:

```
scale * rotation * translation
```

The order these operations take place is from right to left. Therefore, translation is multiplied by rotation, and then the result is multiplied by scale. In our code, if you take a look at our matrix multiplication, we are already applying the transformation matrix on the right of the operation and, as such, achieve a different order. Again, if you are ever in doubt if your code is working as it should, double-check it via different calculation methods including using online calculators.

Having the transformation of an object in a neat matrix is a nice feature, but there will come times, for example, when you are animating, when you need to obtain the current position, rotation, and scale of an object. As we have already discussed, finding the translation or position of an object is achieved by extracting the last column of the matrix. But how do we find rotation and scaling? Let's start with the transformation matrix:

$$\begin{bmatrix} a & b & c & d \\ e & f & g & h \\ i & j & k & l \\ 0 & 0 & 0 & 1 \end{bmatrix}$$

The position in this matrix is the (d, h, l) vector.

The scale is the length of the first three column vectors, such that the x scale is the length of (a, e, i), the y scale is the length of (b, f, j), and the z scale is the length of (c, g, k). We can use Pythagoras' theorem, as discussed in *Chapter 9, Practicing Vector Essentials*, to calculate these lengths.

The rotation matrix is what remains after the scaling and translations are taken out. So, for each column, the values are divided by their respective scale, like this:

$$\begin{bmatrix} a/s_x & b/s_y & c/s_z & 0 \\ e/s_x & f/s_y & g/s_z & 0 \\ i/s_x & j/s_y & k/s_z & 0 \\ 0 & 0 & 0 & 1 \end{bmatrix}$$

I'll leave the implementation of this as an exercise for you to work on by yourself before continuing.

Your turn...

Exercise A. Create three new methods in the `Transform` class called `get_position()`, `get_scale()`, and `get_rotation()` that return the values associated with each by extracting them from the modelview matrix.

In this section, we have examined the modelview matrix and recreated and applied it in our OpenGL project. You might be asking yourself why we would bother with such intricacies when there are OpenGL API methods that allow us to interact with the modelview matrix in a much simpler way. The answer is twofold. First, it is important that you understand the nature and construction of the modelview matrix to take your object manipulation skills to the next level, and second, you should possess the skills to program your own modelview matrix in the event you find yourself in a situation where such OpenGL methods aren't available.

One such situation is in transitioning from the version of OpenGL we are using in this book to the vertex/fragment shader methods implemented in OpenGL 3.x, which no longer uses the `glTranslate`, `glRotate`, and `glScale` methods. However, at this stage, it is critical for your learning journey that you have an appreciation of these methods and their equivalent matrix operations before transitioning to vertex/fragment shader modeling in later chapters.

With knowledge of the modelview matrix in your graphics toolkit, it's now time to examine the view matrix.

Working with the View Matrix

The view matrix takes the world space coordinates produced by the modelview matrix and transforms them into the camera or eye space. The eye space assumes the origin of this space to be at the position of the camera or viewer's eye. As we saw in *Chapter 4, Graphics and Game Engine Components*, the eye space can either be a frustum (a pyramid with the top cut off for perspective views) or orthogonal (a rectangular prism for parallel views). The view matrix can take translation and rotations like the modelview matrix, but instead of transforming individual objects, it transforms everything in the world. That's because it's basically the equivalent of moving a camera around in the world, and then determining what the world will look like through that camera.

Unlike the modelview matrix, which can be loaded after an OpenGL call to `glGetDoublev( GL_ MODELVIEW_MATRIX)`, the view matrix sets a special matrix mode. The view matrix is multiplied with the modelview matrix, and then the resulting combined matrix is applied to any geometry in the environment. This geometry is made up of all the points representing any vertices of meshes present in the 3D scene.

The easiest way to see how this works is to implement a `Camera` class in our project that will allow us to move a camera around in the scene.

Let's do it...

In this exercise, we will create a `Camera` class that can maintain the location of the camera viewing the world as well as enable it to move around the scene. Follow these steps:

1. Create a new Python script called `Camera.py` and add the following code to it:

```
import pygame
import math
import numpy as np

class Camera:
    def __init__(self):
        self.VM = np.identity(4)

    def get_VM(self):
        return self.VM
```

The Camera class maintains a 4x4 matrix that is initialized to an identity matrix. This places the camera at (0, 0, 0) in the world and has it facing down the positive direction along the z axis. The class also provides a way of returning this matrix with the get_VM() method to return the view matrix for use elsewhere in the project.

The get_position() and update_position() methods work together in the same way they did for the Transform class:

```
def get_position(self):
    position = pygame.Vector3(self.VM[0, 3],
                              self.VM[1, 3],
                              self.VM[2, 3])

    return position

def update_position(self, position:
                    pygame.Vector3):
    self.VM = self.VM @ np.matrix([[1, 0, 0, 0],
                                   [0, 1, 0, 0],
                                   [0, 0, 1, 0],
                                   [position.x,
                                    position.y,
                                    position.z,
                                    1]])
```

The `get_position()` method returns the position of the camera from the final row in the view matrix, and this is used by the `update_position()` method to update the camera's position by translating it by the vector value passed through. The position is stored in the view matrix.

2. To update the position of the camera, we are adding a very simple `update()` method that we can call from our main program to allow forward and backward movement of the camera along the *z* axis with the *W* and *S* keys:

```
def update(self):
    key = pygame.key.get_pressed()
    if key[pygame.K_w]:
        self.update_position(self.get_position() +
                                        pygame.Vector3(0, 0,
                                                       0.01))

    if key[pygame.K_s]:
        self.update_position(self.get_position() +
                                        pygame.Vector3(0, 0,
                                                       -0.01))
```

3. In `TransformationMatrices.py`, instantiate the camera thus:

```
from OpenGL.GLU import *
from Cube import *
from Camera import *

..

trans.update_scale(pygame.Vector3(0.5, 2, 1))
trans.rotate_x(45)
trans.update_position(pygame.Vector3(0, 0, -3))

camera = Camera()

objects_3d.append(cube)

clock = pygame.time.Clock()

..
```

4. Inside the `while` loop in `TransformationMatrices.py`, add a line to call the camera's update, and then pass the camera to the object update, like this:

```
while not done:
    events = pygame.event.get()
    ..


    glPushMatrix()
    glClear(GL_COLOR_BUFFER_BIT | GL_DEPTH_BUFFER_BIT)
    camera.update()
    set_3d()


    for o in objects_3d:
        o.update(camera, events)


    set_2d()
..
```

5. Finally, in `Object.py`, you can modify the `update()` method to multiply the modelview matrix with the camera's view matrix to allow the camera's movement to modify the way the scene is viewed thus:

```
..
from Camera import *
..
def update(self, camera: Camera, events = None):
    glPushMatrix()
    for c in self.components:
        if isinstance(c, Transform):
            glLoadMatrixf(c.get_MVM() *
                          camera.get_VM())
            #mv = glGetDoublev(GL_MODELVIEW_MATRIX)
            #print("MV: ")
            #print(mv)
```

Note that at this point, you can remove or comment out the lines printing the modelview matrix unless, of course, you are interested in what it contains.

6. Run `TransformationMatrices.py`. You will now be able to move the camera using the w key to move toward the cube and the s key to move away. Without modifying the transform that belongs to the cube, the scene can be changed to allow for the camera's viewing location, as shown in *Figure 14.3*:

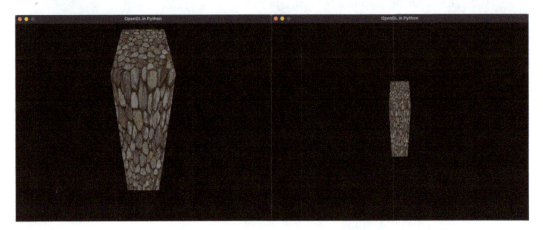

Figure 14.3: Moving toward and away from the cube

In this exercise, we started creating a `Camera` class to allow for the movement of the camera around the world. Currently, it only moves forward and backward; however, we will expand on this in *Chapter 15*, *Navigating the View Space*, to allow sideways movement and rotations.

In this section, we have investigated the view matrix. As it gets blended with the modelview matrix in the same step, it might seem like a superfluous operation. However, the modelview matrix is one that each graphics object maintains as its individual transform, whereas the view matrix is influenced by camera movement and applied to all objects in the world.

The final matrix we need to examine to complete our discussion on the OpenGL matrix stack is the projection matrix.

Working with the Projection Matrix

The objective of the projection matrix is to take objects from view space into projection space. Basically, this takes coordinates inside the camera's viewing volume (specified by the shape and size of the frustum or rectangular prism) and puts them in **normalized device coordinates** (**NDCs**). The calculations for a perspective view's projection matrix were explained in *Chapter 4*, *Graphics and Game Engine Components*. The OpenGL documentation also specifies how it constructs the projection matrix with a call to `gluPerspective()` at `registry.khronos.org/OpenGL-Refpages/gl2.1/xhtml/gluPerspective.xml` where the mathematics is revealed. The projection matrix is defined as follows:

$$\begin{bmatrix} a & 0 & 0 & 0 \\ 0 & b & 0 & 0 \\ 0 & 0 & c & d \\ 0 & 0 & -1 & 1 \end{bmatrix}$$

Here, the values of *a*, *b*, *c*, and *d* are calculated from the parameters passed to `gluPerspective()`, which are the vertical **field of view** (fovy), the aspect ratio of the window, the near plane, and the far plane. These values are determined thus:

f = *cotangent(fovy/2)* = *1/tan(fovy/2)*

a = *f/aspect ratio*

b = *f*

c = *(far + near)/(near – far)*

d = *(2 * near * far) /(near – far)*

From the positions in the matrix of the calculations for a, b, c, and d, you can see that a, b, and c are scaling values for the *x*, *y*, and *z* dimensions respectively, and d is the *z*-axis offset. Take a look at the calculations for each of these, and you will be able to determine whether this makes sense, as you are wanting to take the world's camera view and rescale and move it into NDC coordinates.

OpenGL sets the projection matrix when in GL_PROJECTION matrix mode, and in our project, we've used the `gluOrtho2D()` method to set a parallel projection and the `gluPerspective()` method to set a perspective projection. Both of these can be examined in the `set_2D()` and `set_3d()` methods in our project. As with `glTranslate()` and other transformation methods, these projection matrix creation methods are also obsolete in OpenGL 3.x, and therefore it's worth examining these in detail here.

Let's now replace the call to `gluPerspective()` in our project with a projection matrix we construct.

Let's do it...

In this exercise, we will use the `Camera` class to store its own projection matrix that can be used in the GL_PROJECTION_MATRIX mode. Follow these steps:

1. Modify the `Camera` class to include the calculations to create a projection matrix, like this:

```
class Camera:
    def __init__(self, fovy, aspect, near, far):
        f = 1/math.tan(math.radians(fovy/2))
        a = f/aspect
        b = f
```

```python
        c = (far + near) / (near - far)
        d = 2 * near * far / (near - far)
        self.PPM = np.matrix([
            [a, 0, 0, 0],
            [0, b, 0, 0],
            [0, 0, c, -1],
            [0, 0, d, 0]
        ])
        self.VM = np.identity(4)

    def get_VM(self):
        return self.VM

    def get_PPM(self):
        return self.PPM
```

Here, we are defining a **perspective projection matrix (PPM)**. The calculations for each component of the matrix are taken from the OpenGL definition for `gluPerspective()`. Note the positions of d and -1 in the matrix are swapped to give the inverted matrix because of the way we are multiplying them at the end of the PPM instead of in front. A method called `get_PPM()` has been added to return the matrix for use in the main program. Here, you should also note how we will be accepting the `fovy` value in degrees as does the `gluPerspective()` method, but because of the math functions in Python, we need to ensure these degrees are converted to radians before any calculations are performed.

2. In the `TransformationMatrices.py` script, modify the creation of the camera to the following:

```python
..
trans.rotate_x(45)
trans.update_position(pygame.Vector3(0, 0, -3))

camera = Camera(60, (screen_width / screen_height),
                0.1,
                1000.0)

objects_3d.append(cube)
```

```
clock = pygame.time.Clock()
..
```

3. Still in the `TransformationMatrices.py` script, in the `set_3d()` method, use the camera's projection matrix to set the projection mode instead of `gluPerspective()`, like this:

```
def set_3d():
    glMatrixMode(GL_PROJECTION)
    glLoadMatrixf(camera.get_PPM())
    #glLoadIdentity()
    #gluPerspective(60, (screen_width /
    #                screen_height), 0.1,
    #                1000.0)
    pv = glGetDoublev(GL_PROJECTION_MATRIX)
    print("PV: ")
    print(pv)
    glMatrixMode(GL_MODELVIEW)
    glLoadIdentity()
..
```

Note that I have also included code to print out the projection matrix so that you can compare the results of both techniques if you are interested.

4. Run the project. Changing to the use of our own projection matrix will not change how your project runs, and you will still be able to move the camera with the w and s keys.

In this exercise, we've given the world camera a projection matrix that deals with the perspective projection of objects in the world. In the project, we also have a 2D view with a parallel camera controlled by a call to the `gluOrtho2D()` method. Just as `gluPerspective()` has an associated matrix, so does `gluOrtho2D()`. The next challenge for you is to implement this.

Your turn...

Exercise B. In your Python project, create a `Camera2D` class that stores a parallel projection matrix. Use the matrix you create in the new `Camera2D` class to replace the call to `gluOrtho2D()` in the `set_2d()` method in your main program. The documentation for `gluOrtho2D()` can be found at `https://registry.khronos.org/OpenGL-Refpages/gl2.1/xhtml/gluOrtho2D.xml`. You will also require the documentation for `glOrtho()`, which can be found here: `https://registry.khronos.org/OpenGL-Refpages/gl2.1/xhtml/glOrtho.xml`.

The projection matrix is the final piece of the puzzle when converting model coordinates into screen coordinates. Together with the modelview matrix and the view matrix, they contain all the transformational information to render a vertex in the correct position on a computer screen.

Summary

In this chapter, we have explored the principal modelview, view, and projection matrices that comprise the OpenGL matrix stack. As discussed, these matrices are key to understanding how model coordinates are transformed into screen positions for the rendering of objects. No matter which graphics engine you work with, these matrices are always present. As you move into vertex/fragment shading, it is key that you understand the role each of these matrices plays, and the order they are applied to model coordinates.

In the next chapter, we will work further to improve the maneuverability of the camera that we created herein by adding advanced movement and rotational abilities. This will allow you to fly the camera through the scene and orient it for viewing from any angle.

Answers

Exercise A.

```
def get_position(self):
    position = pygame.Vector3(self.MVM[0, 3],
                              self.MVM[1, 3],
                              self.MVM[2, 3])
    return position


def get_scale(self):
    sx = pygame.Vector3(self.MVM[0, 0], self.MVM[1, 0],
                        self.MVM[2, 0])
    sy = pygame.Vector3(self.MVM[0, 1], self.MVM[1, 1],
                        self.MVM[2, 1])
    sz = pygame.Vector3(self.MVM[0, 2], self.MVM[1, 2],
                        self.MVM[2, 2])
    return pygame.Vector3(sx.magnitude(), sy.magnitude(),
                          sz.magnitude())


def get_rotation(self):
    scale = self.get_scale()
```

```
rotation = np.identity(4)
rotation[0, 0] = self.MVM[0, 0] / scale.x
rotation[0, 1] = self.MVM[0, 1] / scale.x
rotation[0, 2] = self.MVM[0, 2] / scale.x
rotation[1, 0] = self.MVM[1, 0] / scale.y
rotation[1, 1] = self.MVM[1, 1] / scale.y
rotation[1, 2] = self.MVM[1, 2] / scale.y
rotation[2, 0] = self.MVM[2, 0] / scale.z
rotation[2, 1] = self.MVM[2, 1] / scale.z
rotation[2, 2] = self.MVM[2, 2] / scale.z
```

Exercise B.

Reading the documentation for `gluOrtho2D()` will reveal that it uses the same projection matrix as `glOrtho()` but with the near plane value set to -1 and the far plane value set to 1. Creating a 2D camera class for this involves performing the relevant calculations and creating a 4x4 projection matrix. This class doesn't require code to move it as it's being used for the user interface view. The code for `Camera2D.py` is, therefore, this:

```python
import numpy as np


class Camera2D:
    def __init__(self, left, right, top, bottom):
        near_val = -1
        far_val = 1
        a = 2/(right-left)
        b = 2/(top-bottom)
        c = -2/(far_val - near_val)
        d = -(right + left)/(right - left)
        e = -(top + bottom)/(top - bottom)
        f = -(far_val + near_val)/(far_val - near_val)
        self.PPM = np.matrix([
            [a, 0, 0, 0],
            [0, b, 0, 0],
            [0, 0, c, 0],
            [d, e, f, 0]
        ])
```

```
            self.VM = np.identity(4)


    def get_PPM(self):
        return self.PPM
```

With this completed, you will update the `TransformationMatrices.py` script to use this camera thus:

```
..
from Cube import *
from Camera import *
from Camera2D import *


pygame.init()
..


trans.update_position(pygame.Vector3(0, 0, -3))


camera = Camera(60, (screen_width / screen_height), 0.1,
1000.0)
camera2D = Camera2D(gui_dimensions[0], gui_dimensions[1],
                    gui_dimensions[3], gui_dimensions[2])


objects_3d.append(cube)


..


def set_2d():
    glMatrixMode(GL_PROJECTION)
    glLoadMatrixf(camera2D.get_PPM())
    #glLoadIdentity()  # reset projection matrix
    #gluOrtho2D(gui_dimensions[0], gui_dimensions[1],
```

```
                    #gui_dimensions[3], gui_dimensions[2])

    glMatrixMode(GL_MODELVIEW)


..
```

To confirm your matrix is correct, you can print it out and compare it with the one created by `gluOrtho2D()`, as I suggested you do with the perspective camera.

## 15

Navigating the View Space

One of the most common uses of transformations in computer games is the movement of the camera viewing the scene. While the camera can be considered as another 3D object in the environment with respect to its movement and orientation, its current transformation is also used as the view matrix by which all virtual objects in the world are affected.

The transformations presented in this chapter for moving and rotating the camera within its environment are the exact same operations that would be performed on any object in the environment.

In this chapter we will explore:

- Flying maneuvers
- Understanding and fixing compound rotation quirks
- Improving camera orientations

By the end of this chapter, you will have a greater appreciation of the complexity of 3D rotations and the issues that arise from using them in mathematical operations to combine multiple orientations. The knowledge herein you will use over and over again as you work in graphics and games as the concepts are fundamental elements of all virtual environments in which you need to move objects or the camera.

Technical requirements

In this chapter, we will be using the Python, PyCharm and Pygame as used in previous chapters.

Before you begin coding, create a new folder in the PyCharm project for the contents of this chapter called Chapter_15.

The solution files containing the code can be found on GitHub at https://github.com/PacktPublishing/Mathematics-for-Game-Programming-and-Computer-Graphics/tree/main/Chapter15.

Flying maneuvers

Any object in 3D space can have transformations applied to it. It's no different for the camera. In this section, we will explore how the same mathematics can be applied to the object to move and reorient it.

In *Chapter 12, Mastering Affine Transformations*, we discussed the three rotations that can take place in 3D space; namely, **pitching, yawing,** and **rolling**. These correspond with rotations around the x, y, and z axes respectively.

The 4 x 4 matrix that will perform a pitch around the x axis is:

$$\begin{bmatrix} 1 & 0 & 0 & 0 \\ 0 & \cos(\theta) & \sin(\theta) & 0 \\ 0 & -\sin(\theta) & \cos(\theta) & 0 \\ 0 & 0 & 0 & 1 \end{bmatrix}$$

The matrix that will perform a yaw around the y axis is:

$$\begin{bmatrix} \cos(\theta) & 0 & \sin(\theta) & 0 \\ 0 & 1 & 0 & 0 \\ -\sin(\theta) & 0 & \cos(\theta) & 0 \\ 0 & 0 & 0 & 1 \end{bmatrix}$$

The matrix to perform a roll around the z axis is:

$$\begin{bmatrix} \cos(\theta) & \sin(\theta) & 0 & 0 \\ -\sin(\theta) & \cos(\theta) & 0 & 0 \\ 0 & 0 & 1 & 0 \\ 0 & 0 & 0 & 1 \end{bmatrix}$$

We can use these matrices to rotate the camera and thus the view space. To do this, we multiply the view matrix by one or more of these.

The way the camera, or any other object transformed with these matrices, rotations can be likened to the movement of an aircraft as illustrated in *Figure 15.1*. In *Figure 15.1 (a)*, the rotational movement around each axis is shown. In *(b)*, a pitch rotation shows how the nose of the plane looks up and down pivoting around an x axis that extends to the left and right of the plane. Then *(c)* demonstrates how a roll tilts the plane from side to side and *(d)* demonstrates how a yaw rotates the plane on the spot around its up axis. This is how a first-person character in a game environment looks around:

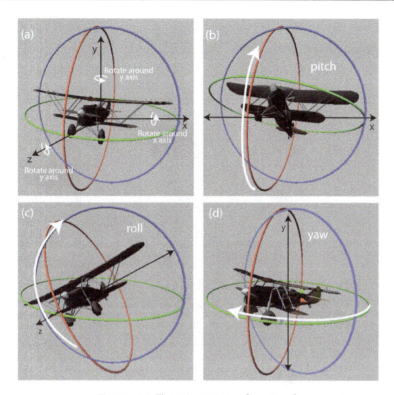

Figure 15.1: The orientations of an aircraft

It is useful in 3D graphics environments to be able to orient the camera around the scene using these rotations. This is how a first-person character in a game environment looks around. Of course, the best way to experience these maneuvers is to put them into practice as we will do now.

Let's do it...

In this practical exercise, we will implement roll, pitch, and yaw in the `Camera.py` class to enable you to move the camera around a scene:

1. Make a copy of the `Chapter_14` folder and name it `Chapter_15`.

    The rotation matrices we need for the camera are in the `Transform` class. In fact, all the operations required for matrix manipulation to move and orient the camera are in the `Transform` class, and therefore rather than repeat code, it makes more sense to give the camera a transform component that can do all the work.

2. To achieve this, go ahead and modify the `Camera` class as follows:

    ```
    import pygame
    import math
    ```

```python
import numpy as np
from Transform import *


class Camera:
    def __init__(self, fovy, aspect, near, far):
        ..
        self.PPM = np.matrix([
            [a, 0, 0, 0],
            [0, b, 0, 0],
            [0, 0, c, -1],
            [0, 0, d, 0]
        ])
        self.VM = np.identity(4)
        self.transform = Transform()
        self.pan_speed = 0.01

    def get_VM(self):
        return self.transform.MVM

    def get_PPM(self):
        return self.PPM

    def update(self):
        key = pygame.key.get_pressed()
        if key[pygame.K_w]:
            self.transform.update_position(
                self.transform.get_position()
                + pygame.Vector3(0, 0, self.pan_speed),
                False)
        if key[pygame.K_s]:
            self.transform.update_position(
                self.transform.get_position()
                + pygame.Vector3(0, 0,
                -self.pan_speed), False)
```

The key things to notice in this class are the addition of the transform property in the initializer and then its use to call the update_position() method in the update() method. Now that the Camera class is using the transform to store and manipulate the view matrix, in what's called the MVM (ModelView matrix) in the Transform class, the get_VM() method is updated to return the MVM from the transform. The model view matrix in the Transform class is just a 4x4 matrix and you could say it holds the model view matrix that would be local to the camera itself (if the camera were just another object in the scene). Hence, we can use that matrix belonging to the camera as a view matrix. This is particularly useful as you might want to swap view matrices between a multitude of cameras. The camera's new transform.MVM only becomes the view matrix if it is the camera being used to view the scene.

We are also passing a False parameter through the update_position() method, which we will now cater for.

3. Because the movement of the camera means that the oppositive movement is added to any game objects in world coordinates, we must detail with its position update in the opposite order. To Transform.py make the following modifications:

```
def update_position(self, position: pygame.Vector3,
                    local=True):
    if local:
        self.MVM = self.MVM @ np.matrix([[1, 0, 0, 0],
                                         [0, 1, 0, 0],
                                         [0, 0, 1, 0],
                                         [position.x,
                                          position.y,
                                          position.z,
                                          1]])

    else:
        self.MVM = np.matrix([[1, 0, 0, 0],
                              [0, 1, 0, 0],
                              [0, 0, 1, 0],
                              [position.x, position.y,
                               position.z, 1]])
                    @ self.MVM
```

Notice how the matrix multiplication is occurring in the opposite order if local is given a value of False. This is because we want to move the camera in world space and apply that to the entire scene.

With the modifications made to the camera, you can run the project from the main script file: `TransformationMatrices.py` and the scene will be as they were before where the camera can be moved in and out with the '*W*' and '*S*' keys.

Now as a challenge, let's see whether you can program in the movements for the '*A*' and '*D*' keys.

Your turn...

Exercise A. Give the camera the ability to pan (slide) to the left when the '*A*' key is pressed and to pan to the right when the '*D*' key is pressed and to yaw when the '*Q*' and '*E*' keys are used.

1. To make the scene more interesting to explore with the camera, I have made a low-polygon version of the teapot model called `teapotSM.obj`. You can download this from *GitHub* and add this to your `models` folder for the project and then load it in the main script of `TransformationMatrices.py` like this:

    ```
    from LoadMesh import *
    objects_2d = []


    teapot = Object("Teapot")
    teapot.add_component(Transform())
    teapot.add_component(LoadMesh(GL_LINE_LOOP,
                         "models/teapotSM.obj"))
    trans: Transform = teapot.get_component(Transform)


    trans.rotate_y(90)
    trans.update_position(pygame.Vector3(0, -2, -3))


    camera = Camera(60, (screen_width / screen_height),
                    0.1, 1000.0)
    objects_3d.append(teapot)
    clock = pygame.time.Clock()
    ```

2. If you still have print lines in your code printing out matrices in the console, such as the ones in the `update()` method of `Object.py`, you can remove them if you desire.

    When first run, you'll find a wireframe image of a teapot as shown in *Figure 15.2*. If you want of move the camera faster, increase the value of `pan_speed` in `Camera.py`.

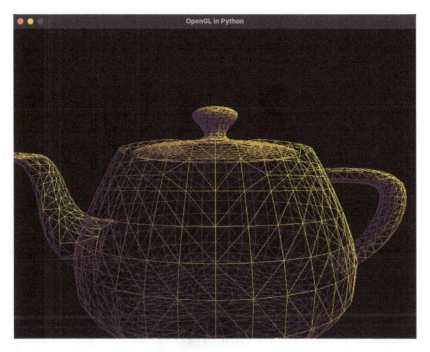

Figure 15.2: Low-polygon teapot

3. To color the teapot as in *Figure 15.2*, add some lights into the `setup_3d()` method as follows:

```
glEnable(GL_LIGHTING)
glLight(GL_LIGHT0, GL_POSITION, (5, 5, 5, 0))
glLightfv(GL_LIGHT0, GL_AMBIENT, (1, 0, 1, 1))
glLightfv(GL_LIGHT0, GL_DIFFUSE, (1, 1, 0, 1))
glLightfv(GL_LIGHT0, GL_SPECULAR, (0, 1, 0, 1))
glEnable(GL_LIGHT0)
```

4. Now it's time to add rotations to the camera. For this, we will first explore a yaw using the '*Q*' and '*E*' keys. To do this, add the following code to `Camera.py`:

```
def __init__(self, fovy, aspect, near, far):
    ..
    self.transform = Transform()
    self.pan_speed = 0.1
    self.rotate_speed = 10
..
def update(self):
```

```
key = pygame.key.get_pressed()
..
if key[pygame.K_d]:
self.transform.update_position(
            self.transform.get_position()
            + pygame.Vector3(-self.pan_speed, 0,
            0))
if key[pygame.K_q]:
    self.transform.rotate_y(self.rotate_speed)
if key[pygame.K_e]:
    self.transform.rotate_y(-self.rotate_speed)
```

In the new code, we have added a property to store the rotation speed, which you can change to rotate faster or slower, and key press tests for *Q* and *E*, which will make use of the `rotate_y()` method we wrote earlier.

5. Run the program now and you will find that you can rotate the camera to the left and right around its *y* axis. If you would prefer the keys rotated in the opposite direction to the way I have them, you can remove the minus sign from the '*E*' press line and add it to the *Q* press line.

    From here the code to pitch and roll the camera is elementary. All you need to do is program extra keypresses and call the `rotate_x()` and `rotate_z()` methods. Try this for yourself! However, now we are going to continue by programming the mouse to rotate the camera.

6. Before we continue, make a copy of `TransformationMatrices.py` and call it `FlyCamera.py`. Near the `while` loop you should make these changes to use the mouse:

```
pygame.event.set_grab(True)
pygame.mouse.set_visible(False)
while not done:
    events = pygame.event.get()
    for event in events:
        if event.type == pygame.QUIT:
            done = True
        if event.type == KEYDOWN:
            if event.key == K_ESCAPE:
                pygame.mouse.set_visible(True)
                pygame.event.set_grab(False)
            if event.key == K_SPACE:
                pygame.mouse.set_visible(False)
```

```
                    pygame.event.set_grab(True)
```

```
    glPushMatrix()
```

The first new lines use pygame to grab the mouse so it can be used to feed motion into the program. After that the mouse's visibility is set to `False` so it is no longer visible in the window. Because at some time you might like to get the mouse back and stop it having motion control and switch back again, a `KEYDOWN` event is monitored. If the key is pressed, the mouse becomes visible and you can move it freely. This is very useful if you want to close the window or use in for any user interface elements such as buttons. However, if you want to go back into mousegrabbing mode, the spacebar can be used.

7. To read the mouse movement and use the changes to rotate the camera, modify the `Camera.py` script like this:

```
    def __init__(self, fovy, aspect, near, far):
        ..
        self.rotate_speed = 10
        self.last_mouse = pygame.math.Vector2(0, 0)
        self.mouse_sensitivityX = 0.5
        self.mouse_sensitivityY = 0.5
        self.mouse_invert = -1
```

In the initialization method we add four new properties. `last_mouse` stores the last read position of the mouse. This is required for us to calculate how far the mouse has moved since the last frame. This value can then be used as a rotation amount. Following this, the mouse sensitivity is set for movements in *x* (horizontally across the screen) and *y* (vertically up and down the screen). And last, an invert setting is added in case you prefer the mouse movements to be in the opposite direction to which you are scrolling. This is a common optional setting in first-person shooter and flying games.

Next, in the `update()` method the mouse movements are captured and used to rotate the camera if the mouse is in the invisible mode set in the `FlyCamera.py` loop in *step 5*. Here, the mouse's position is captured and taken away from the position of the mouse in the last frame. This gives the total movement of the mouse during the last frame:

```
def update(self):
    key = pygame.key.get_pressed()
    if key[pygame.K_w]:
    ..

            if key[pygame.K_e]:
                self.transform.rotate_y(
                    -self.rotate_speed)
```

```
if not pygame.mouse.get_visible():
    mouse_pos = pygame.mouse.get_pos()
    mouse_change = self.last_mouse -
                pygame.math.Vector2(mouse_pos)
    pygame.mouse.set_pos(
            pygame.display.get_window_size()[0]
            / 2,
            pygame.display.get_window_size()[1]
            / 2)
    self.last_mouse = pygame.mouse.get_pos()
    self.rotate_with_mouse(mouse_change.x *
        self.mouse_sensitivityX,
        mouse_change.y *
    self.mouse_sensitivityY)
```

Following this the position of the mouse is set to the center of the window using values stored by pygame. Even though the mouse can't be seen, it's still possible to scroll it outside the window. This would result in incorrect measurements of how far the mouse has moved. Therefore, for each frame, it is set back to the center. The next line of code resets the `last_mouse` value, which should be the center of the screen at this point.

Finally, the values recorded for the mouse's change in position are used to rotate the camera inside a new method we will add next.

8. To the `Camera` class add the following method:

```
def rotate_with_mouse(self, yaw, pitch):
        self.transform.rotate_y(self.mouse_invert * yaw *
            self.mouse_sensitivityY)
        self.transform.rotate_x(self.mouse_invert *
            pitch * self.mouse_sensitivityX)
```

This method will rotate the mouse around the *y* axis using the mouse's vertical movement and around the *x* axis using the mouse's horizontal movement.

You can now run the program to test out the mouse movement. Note if it is too fast, turn down the mouse sensitivity values. You will also be able to use the keys for moving the camera in addition to turning it. At this point, it won't work perfectly but it is a start.

In this exercise, we added code to move the camera around in its 3D environment as well as rotate it. The biggest issue you will have experienced is that the camera starts to tip roll around the z axis and may eventually turn upside down. This can easily be achieved by making small circular motions with the mouse. This might seem counterintuitive since you are only coding it to pitch and yaw. The reason why this is happening and a solution to the problem is the topic of the next section.

Understanding and fixing compound rotation quirks

In the previous section, you had the chance to experiment with moving the camera around in its 3D environment so you could explore looking at the environment from different angles. While moving the camera is elementary, adding rotations seems to introduce undesired results. In this section, you will discover the source of these issues and look at how to fix them.

Imagine you are holding a camera and looking through it toward the horizon. Now bend down at the hip by 90 degrees so that the camera is looking straight at the ground. Next, rotate your upper body 90 degrees upward to the left. You might be thinking along the steps shown in *Figure 15.3*. First rotating around the x axis turns the camera to face downward making its y axis horizontal and then a rotation around the y axis will face the camera sideways and result in the x axis sitting horizontally. In performing these orientations, we assume each axis can move independently of the others:

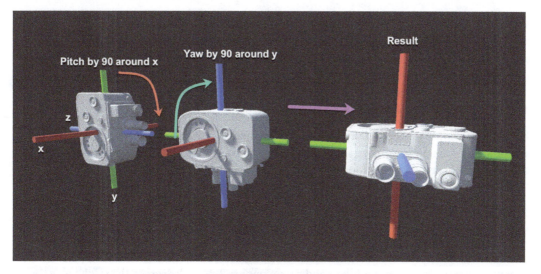

Figure 15.3: Pitching then yawing by 90 degrees

This might be the desired result – however, mathematically this doesn't happen and it's because of the way our matrices represent and calculate rotations. The axes are in fact connected to each other in a hierarchy of gimbals with y on the outside, x next on the inside, and z inside that. Look at *Figure 15.4* where the default gimbal setup is shown on the left.

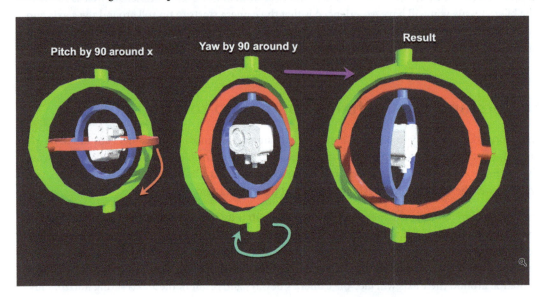

Figure 15.4: Rotating using codependent gimbals

Mathematically, when a 90-degree rotation around the x axis occurs, the x gimbal (shown in red) and the y gimbal (shown in green) become aligned. The x axis is *stuck* in this position such that if a y axis rotation occurs, the x gimbal has no other option than to rotate with it. Each of these gimbals represent a degree of freedom for orientations. In the image on the left there are three degrees of freedom. After an x rotation of 90 though, the x gimbal and y gimbal are aligned into the same dimension and therefore an entire dimension is lost.

To further exemplify this effect, we can look at it mathematically. Here's the generic equation for performing a z rotation, y rotation, and x rotation, using **Euler's** method, with the z rotation matrix being on the right and the x rotation on the left.

$$\begin{bmatrix} 1 & 0 & 0 & 0 \\ 0 & \cos(\phi) & \sin(\phi) & 0 \\ 0 & -\sin(\phi) & \cos(\phi) & 0 \\ 0 & 0 & 0 & 1 \end{bmatrix}\begin{bmatrix} \cos(\theta) & 0 & \sin(\theta) & 0 \\ 0 & 1 & 0 & 0 \\ -\sin(\theta) & 0 & \cos(\theta) & 0 \\ 0 & 0 & 0 & 1 \end{bmatrix}\begin{bmatrix} \cos(\delta) & \sin(\delta) & 0 & 0 \\ -\sin(\delta) & \cos(\delta) & 0 & 0 \\ 0 & 0 & 1 & 0 \\ 0 & 0 & 0 & 1 \end{bmatrix}$$

No matter what the angles used for the x and z rotations are, if θ is 90 degrees, the result will give:

$$\begin{bmatrix} 1 & 0 & 0 & 0 \\ 0 & \cos(\phi) & \sin(\phi) & 0 \\ 0 & -\sin(\phi) & \cos(\phi) & 0 \\ 0 & 0 & 0 & 1 \end{bmatrix} \begin{bmatrix} 0 & 0 & 1 & 0 \\ 0 & 1 & 0 & 0 \\ -1 & 0 & 0 & 0 \\ 0 & 0 & 0 & 1 \end{bmatrix} \begin{bmatrix} \cos(\delta) & \sin(\delta) & 0 & 0 \\ -\sin(\delta) & \cos(\delta) & 0 & 0 \\ 0 & 0 & 1 & 0 \\ 0 & 0 & 0 & 1 \end{bmatrix}$$

Multiplying this out results in the compound matrix:

$$\begin{bmatrix} 0 & 0 & 1 & 0 \\ \sin(\phi)\cos(\delta) + \cos(\phi)\sin(\delta) & -\sin(\phi)\sin(\delta) + \cos(\phi)\cos(\delta) & 0 & 0 \\ -\cos(\phi)\cos(\delta) + \sin(\phi)\sin(\delta) & \cos(\phi)\sin(\delta) + \sin(\phi)\cos(\delta) & 1 & 0 \\ 0 & 0 & 0 & 1 \end{bmatrix}$$

Without knowing the values for the other angles, the resulting matrix tells us that any rotation we wanted to include for the angle θ has been totally lost and therefore so has a dimension of freedom. This effect occurs in many instances of 90-degree rotations that can affect the gimbals. It's known as gimbal lock and is a common issue with gimbal-based rotation systems as is documented as an issue with the famous *Apollo 13* mission to *The Moon* (`https://history.nasa.gov/afj/ap13fj/12day4-approach-moon.html`).

In this section we have explored how compound rotation mathematics we've implemented can break down. While it can work in many situations, the inherent flaw in Euler's method can, at times, accidentally eliminate a degree of rotational freedom. Therefore, a better approach to working with compound rotations requiers a different form of rotational mathematics. It involves representing a rotation as vectors and angles in a format called a **Quaternion**. This is a very advanced mathematical concept that will be explored in the next chapter. For now though, we will fix our projects camera by restricting its rotations and making them far more intuitive with respect to a 3D flying camera.

Improving camera orientations

Regardless of the direction the camera is facing, when flying around in a 3D environment, being able to yaw around the world's up axis is far less disorientating than yawing at an unexpected angle, no matter how mathematically correct it may be. Rolling around the forward-facing vector of the camera is however intuitive as it is the way the viewer is facing. Many first-person player controllers are set up to rotate around the world's up axis and the camera-forward axis as shown in *Figure 15.5*. It prevents the camera tipping accidentally upside down and experiencing gimbal lock.

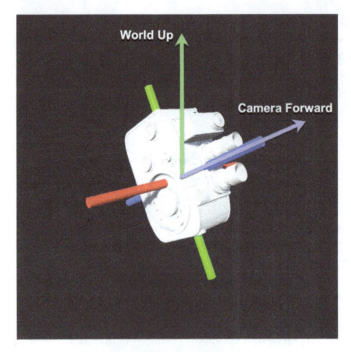

Figure 15.5: Intuitive camera flying rotation axes

To instigate rotations using these axes, we need to be able to switch between local and world space. In this instance we want to rotate the camera in world space (where it yaws around the up axis) and roll around the local *z* axis, which is the axis that indicates the direction the camera is facing. Let's now update the project to use this method of rotation.

Let's do it...

In this exercise, we will modify the `Transform` class to work with local and world position systems and then modify the `Camera` class to rotate like a traditional first-person camera.

1. The modifications to the rotation methods in the `Transform` class require the addition of a new parameter like this:

```python
def rotate_x(self, amount, local=True):
    amount = math.radians(amount)
    if local:
        self.MVM = self.MVM @ np.matrix([
            [1, 0, 0, 0],
            [0, math.cos(amount), math.sin(amount),
            0],
```

```
                [0, -math.sin(amount), math.cos(amount),
                 0],
                [0, 0, 0, 1]])
        else:
            self.MVM = np.matrix([
                [1, 0, 0, 0],
                [0, math.cos(amount), math.sin(amount),
                 0],
                [0, -math.sin(amount), math.cos(amount),
                 0],
                [0, 0, 0, 1]]) @ self.MVM


    def rotate_y(self, amount, local=True):
        amount = math.radians(amount)
        if local:
            self.MVM = self.MVM @ np.matrix([
                [math.cos(amount), 0, -math.sin(amount),
                 0],
                [0, 1, 0, 0],
                [math.sin(amount), 0, math.cos(amount),
                 0],
                [0, 0, 0, 1]])
        else:
            self.MVM = np.matrix([
                [math.cos(amount), 0, -math.sin(amount),
                 0],
                [0, 1, 0, 0],
                [math.sin(amount), 0, math.cos(amount),
                 0],
                [0, 0, 0, 1]]) @ self.MVM


    def rotate_z(self, amount, local=True):
        amount = math.radians(amount)
        if local:
            self.MVM = self.MVM @ np.matrix([
                [math.cos(amount), math.sin(amount), 0,
```

```
                                0],
                          [-math.sin(amount), math.cos(amount), 0,
                                0],
                          [0, 0, 1, 0],
                          [0, 0, 0, 1]])
            else:
                self.MVM = np.matrix([
                          [math.cos(amount), math.sin(amount), 0,
                                0],
                          [-math.sin(amount), math.cos(amount), 0,
                                0],
                          [0, 0, 1, 0],
                          [0, 0, 0, 1]]) @ self.MVM
```

Each of the rotation methods is given an extra parameter to indicate whether the operations should occur locally. Up until now we have been using local rotations and therefore setting this parameter to `True` will ensure the existing code doesn't break.

Next, each one of the methods undergoes the same change. The `if-else` statement checks whether `local` is `True`. If it is, the same matrix multiplication that was performed before is used. If, however, `local` is `False`, the matrix multiplication occurs in reverse order. Performing the operation in reverse is the same as leaving the operation in the same order but inverting the rotation matrix. The inverse of a local matrix, in this case, puts it into world coordinates.

2. Next, change the `rotate_with_mouse()` method to:

```
def rotate_with_mouse(self, yaw, pitch):
    self.transform.rotate_y(self.mouse_invert * yaw *
                                self.mouse_sensitivityY,
False)
    self.transform.rotate_x(self.mouse_invert *
        pitch * self.mouse_sensitivityX, False)
```

The parts of this code doing all the work are the `rotate_y()` method, which operates in world space, and the `rotate_x()` method, which operates in local space.

With these changes made you can play the application. Note, the camera will still be able to go upside down, but it won't end up at a strange angle.

Often with a first-person camera it makes sense to restrict the angles the yaw and roll can travel beyond to stop the camera from going upside down. We therefore need to keep track of the camera's forward vector and the world up vector shown in *Figure 15.6*.

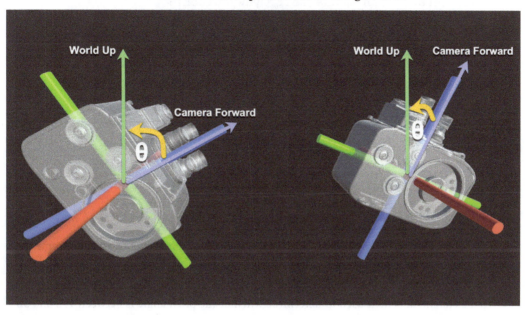

Figure 15.6: Angles between the world up and camera forward

Before we implement the code let's pause to consider the mathematics.

Calculating the camera tilt

At some point in the camera's pitch, you want to be able to stop it from moving any further. This means calculating the angle between the world up vector and the camera's forward vector. As the camera pitches up, this angle gets smaller.

In this case, θ can be calculated as in *Chapter 9, Practicing Vector Essentials,* where we discussed the formula:

$$\cos\theta = \frac{f \cdot g}{|f| \times |g|}$$

There will however be two angles that need to be considered – one where the camera is tilted up and the angle between the up vector and the camera's forward vector is small and another when the camera is tilted down and the angle between the up vector and the camera's forward vector is large. Anywhere in between the camera should be able to tilt freely.

3. Rather than manually coding the entire rotation equations, we can use the built-in methods that come with pygame. Make the following changes to the `rotate_with_mouse()` method:

```
def rotate_with_mouse(self, yaw, pitch):
    forward = pygame.Vector3(self.get_VM()[0, 2],
        self.get_VM()[1, 2], self.get_VM()[2, 2])
    up = pygame.Vector3(0, 1, 0)
    angle = forward.angle_to(up)
    self.transform.rotate_y(self.mouse_invert * yaw *
                            self.mouse_sensitivityY,
                            False)
    if angle < 170.0 and pitch > 0 or //
        angle > 30.0 and pitch < 0:
        self.transform.rotate_x(self.mouse_invert
            * pitch * self.mouse_sensitivityX, True)
```

Here, the forward vector for the camera is being taken from the third column in its transformation matrix, which contains its local *z* axis in world coordinates. Next, the up angle is constructed followed by the use of the pygame `Vector3` class method, `angle_to()`.

An `if` statement is then used to determine whether the pitch angle is between 170 and 30 degrees. If it is within this range, a pitch will still be possible. Note that a restriction of the pitch angle being greater or less than 0 is added as there's no need to rotate around the *x* axis of the camera if there is no pitch angle.

4. Now that you can rotate the camera, you'll notice a small issue that occurs with the movement code. Previously we were using `update_position()` in the `Transform.py` class to add a vector to the position of an object to move it. However, this won't work with an object you want to move in a forward direction no matter which way it is orientated. In *Figure 15.5* the camera has a forward vector, and it is this vector we want to move it along when flying forward. The current code simply adds the world vector of (0, 0, 1), which moves the camera along the world *z* axis. But if the camera isn't facing in this direction, this code won't work. Instead, we need to find the forward-facing vector, as we have for the rotation and use that for movement as well. Modify the code in `Camera.py` like this:

```
def update(self):
    key = pygame.key.get_pressed()
    forward = pygame.Vector3(self.get_VM()[0, 2],
                             self.get_VM()[1, 2],
                             self.get_VM()[2, 2])
    right = pygame.Vector3(self.get_VM()[0, 0],
                           self.get_VM()[1, 0],
```

```
                         self.get_VM()[2, 0])
if key[pygame.K_w]:
    self.transform.update_position(
            self.transform.get_position()
                + forward * self.pan_speed, False)
if key[pygame.K_s]:
    self.transform.update_position(
            self.transform.get_position()
                + forward * -self.pan_speed, False)
if key[pygame.K_a]:
    self.transform.update_position(
            self.transform.get_position()
                + right * self.pan_speed, False)
if key[pygame.K_d]:
    self.transform.update_position(
            self.transform.get_position()
                + right * -self.pan_speed, False)
if key[pygame.K_q]:
    self.transform.rotate_y(self.rotate_speed)
if key[pygame.K_e]:
    self.transform.rotate_y(-self.rotate_speed)
```

Now no matter what direction the camera is facing, it will always move in a consistent manner.

The code for this version of the camera is now complete and you are free to run the project and fly the camera around the teapot.

In this section, we investigated a common method used in first-person controllers in games to prevent the camera from going into gimbal lock as this causes erratic rotations. In addition, code was added to the project to implement this method of rotation as well as restrict the camera from pitching too steeply.

Summary

Rotations would have to be the single most cause of headaches for programmers working in 3D environments. By now you will have an appreciation for the complexity of their mathematics. It's often not until you start using the mathematics that you find out what can go wrong. Only then can you really understand how the errors occur and what you can do to fix them. The issues that affect first-person controller rotations also impact the navigation systems of virtual moving objects such as simulated aircraft and spaceships.

In this chapter, we have explored the mathematics required to move and rotate a virtual camera. This knowledge will allow you to position and angle a camera anywhere in the world to view the objects being rendered. Though the use of several key presses, the camera in your project will now be able to move and rotate. This is essential understanding for any graphics or game programmer, as is being able to eye ball a transformation matrix and have an awareness for the way it will affect the movement of a virtual object.

One thing we did discover in this chapter that has had a monumental impact on graphics mathematics is the limitation of Euler's representation of matrices, which can result in gimbal lock. The solution to removing this limitation is to work with Quaternions, a powerful mathematical concept that confounds the most learned of programmers working with 3D graphics and games. Even though it is an extremely advanced topic, I will give you a gentle introduction to the topic in the next chapter to the end of elucidating the subject matter and showing you how to work with them.

Answers

Exercise A:

In `Camera.py` add extra `if` statements to handle the '*A*' and '*D*' key presses and use the *x* axis to move the camera as follows:

```
def update(self):
    key = pygame.key.get_pressed()
    if key[pygame.K_w]:
        self.transform.update_position(
                self.transform.get_position()
                            + pygame.Vector3(0, 0,
                                            self.pan_speed),
                        False)
    if key[pygame.K_s]:
        self.transform.update_position
                (self.transform.get_position()
                        + pygame.Vector3(0, 0,
                                        -self.pan_speed),
                        False)
    if key[pygame.K_a]:
        self.transform.update_position(
                self.transform.get_position()
                        + pygame.Vector3(self.pan_speed,
                                            0, 0),
```

```
                            False)
if key[pygame.K_d]:
    self.transform.update_position(
        self.transform.get_position()
                    + pygame.Vector3(-self.pan_speed,
                                     0, 0),
                    False)
```

16
Rotating with Quaternions

If you can't remember me saying it before, you'll be sure to hear me say it numerous times throughout this chapter: *quaternions are an advanced mathematical construct*. They are so advanced I don't expect you to fully comprehend them by the end of this chapter. However, what I want you to take away is a healthy appreciation for what they do with respect to solving the gimbal lock issue we discussed in *Chapter 15, Navigating the View Space*.

Besides their usefulness in calculating 3D rotations, quaternions are useful in numerous fields, including computer vision, crystallographic texture analysis, and quantum mechanics. Conceptually, quaternions live in a 4D space through the addition of another dimension to those of the x, y, and z axes used by Euler angles.

In this chapter, we will start with an overview of quaternions and delve into the benefits of their 4D structure. This will reveal how they can be used to replace operations for which we've previously been using Euler rotations. We will then jump back into our Python/OpenGL project and update the rotation methods to use quaternions.

In this chapter, we will be doing the following:

- Introducing quaternions
- Rotating around an arbitrary axis
- Exploring quaternion spaces
- Working with unit quaternions
- Understanding the purpose of normalization

By the end of this chapter, you will have developed the skills to use quaternions in place of Euler angles for compound rotations in graphics applications. Furthermore, you'll have a firm grasp on quaternion concepts and be set up to independently explore the concept in more detail for other applications.

Technical requirements

In this chapter, we will be using Python, PyCharm, and Pygame, as used in previous chapters.

Before you begin coding, create a new folder in the PyCharm project for the contents of this chapter called `Chapter_16`.

The solution files containing the code can be found on GitHub at `https://github.com/PacktPublishing/Mathematics-for-Game-Programming-and-Computer-Graphics/tree/main/Chapter16`.

Introducing quaternions

The minimum number of values needed to represent rotations in 3D space is three. The most intuitive and long-applied method for defining rotations, as we've seen, is to use these values as the three angles of rotation around the x axis, the y axis, and the z axis. The values of these angles can range from 0 to 360 degrees or 0 to 2 PI radians.

Any object in 3D space can be rotated around these axes that represent either the world axes or the object's own local access system. Formally, the angles around the world axes are called **fixed angles**, while the angles around an object's local axis system are called **Euler angles**. However, often both sets of angles are referred to as Euler angles. We covered the mathematics to apply rotations around these three axes in *Chapter 15*, *Navigating the View Space*, in addition to investigating when these calculations break down and cause gimbal lock.

Quaternions were devised in 1843 by Irish mathematician **William Hamilton** for applications to mechanics in 3D space. They can be applied in numerous areas of mathematics, and when Hamilton came up with them, he wasn't even thinking about rotations but rather how to calculate the quotient of two coordinates in 3D space. Put simply, this means being able to perform divisions between coordinates. It's easy enough to add and subtract 3D points and vectors from each other, but how do you multiply and divide them? Hamilton came up with the idea of using four dimensions rather than three. To help you understand the use of four dimensions in a calculation, let's go back to the very beginning of rotational mathematics with those in 2D space.

Rotating around an arbitrary axis

A vector lying on the x axis that is represented by $(1, 0)$ and rotated by θ results in the vector that will be the cosine of the angle and the sine of the angle $(\cos(\theta), \sin(\theta))$ as illustrated in *Figure 16.1*. Likewise, a vector sitting on the y axis represented by $(0, 1)$, when rotated by the same angle, will result in a vector that, too, contains a combination of cosine and sine as $(-\sin(\theta), \cos(\theta))$.

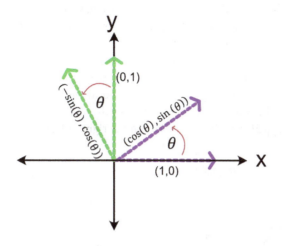

Figure 16.1: Two-dimensional rotations

Do these values look familiar? They should because they are the values we've used in the rotation matrix for a rotation around the *z* axis in *Chapter 15, Navigating the View Space*. Rotating in 2D is essentially the same operation as rotating around the *z* axis; as you can imagine, the *z* axis added to *Figure 16.1* coming out of the screen toward you, and thus rotations in this 2D space are, in fact, rotating around an unseen *z* axis.

All the rotations we've looked at thus far have been to rotate a vector or point around an *x*, *y*, or *z* axis. As for the case of rotating in the 2D space of the x/y plane, which is a rotation around the *z* axis, any of these 3D rotations can be reduced to 2D where a rotation around the *y* axis occurs on the x/z plane, and a rotation around the *x* axis occurs on the y/z plane. This is illustrated in *Figure 16.2*:

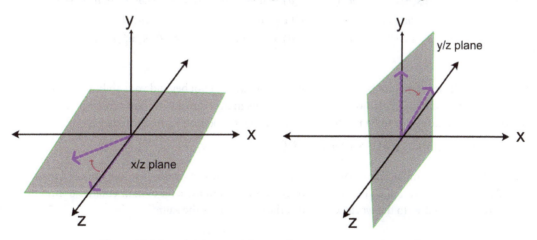

Figure 16.2: Rotations around the z and x axes represented on a plane

Taking this one step further is the **Euler-Rodrigues theorem**, which allows for rotations around an arbitrary vector in space, where this vector becomes the rotational axis. This 3D rotation can also be simplified to a 2D rotation, as shown in *Figure 16.3*. If you consider the 3D rotation of a vector around another vector (u), the tip of the first vector will form a circle in space as it rotates through 360 degrees. This circle represents a plane. Just like the x/y plane in a z rotation. Though the representation of this plane isn't as simple, as we are in a completely different coordinate system that we can't just attribute to straightforward x, y, and z world axes. However, it is still a flat surface and if looked at down the rotational axis, it becomes a 2D rotation, as shown on the right in *Figure 16.2*:

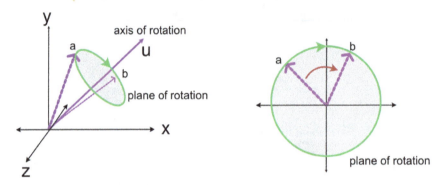

Figure 16.3: Rotations around an arbitrary axis in 3D (left) and reduced to 2D (right)

As previously stated, although we appear to have a nice simple rotation in 2D, because we've moved out of the regular x, y, and z axis system and are now in a different frame of reference, the mathematics becomes somewhat complex. For those of you who are interested, the rotational matrix for rotating around the arbitrary vector u is as follows:

$$\begin{bmatrix} \cos(\theta) + u_x^2(1 - \cos(\theta)) & u_x\,u_y\,(1 - \cos(\theta) - u_z \sin(\theta)) & u_x\,u_z\,(1 - \cos(\theta) + u_y \sin(\theta)) & 0 \\ u_y\,u_x\,(1 - \cos(\theta) + u_z \sin(\theta)) & \cos(\theta) + u_y^2(1 - \cos(\theta)) & u_y\,u_z\,(1 - \cos(\theta) + u_x \sin(\theta)) & 0 \\ u_z\,u_x\,(1 - \cos(\theta) - u_y \sin(\theta)) & u_z\,u_y\,(1 - \cos(\theta) + u_x \sin(\theta)) & \cos(\theta) + u_z^2(1 - \cos(\theta)) & 0 \\ 0 & 0 & 0 & 1 \end{bmatrix}$$

In this format, the mathematics is indeed overwhelming, though not beyond your abilities to program it into your project and use it. However, if we reduce this matrix down to its basic components of the rotational axes u, the vector to be rotated v, and the angle of rotation θ, the formula becomes:

$$R_u(\theta)v = u(u \cdot v) + \cos(\theta)(u \times v) + \sin(\theta)(u \times v)$$

Note here how the familiar terms for $\cos(\theta)$ and $\sin(\theta)$ come out again, just like they did for the original 2D rotations. And that's because although you've had to use a lot of mathematics to find the plane of rotation relative to the world x, y, and z, the operation is the same.

As a reader of this book, I don't expect you to fully embrace the mathematics (unless you want to) as it's university master's level stuff. However, I would like you to appreciate the beauty of it. The main point I'm focusing on is that if you can represent 3D rotations as 2D rotations, then you must be able to represent 4D rotations as 3D rotations. Wrapping your mind around this concept is key to understanding how quaternions work.

Now, you might be wondering if the Rodrigues-Euler Theorem solves the issue of gimbal lock. The answer is, *unfortunately, no*. While the way we've been creating complex rotations with Euler angles thus far allows for compound rotations, the Rodrigues-Euler Theorem does not. Therefore, while they look fabulous, we can't use them in graphics – at least not for complex maneuvers. This is because each rotation that occurs is in a different frame of reference.

Therefore, the solution resides in the four-dimensionality of quaternions, as will soon be revealed.

Exploring quaternion spaces

As we saw in *Chapter 13, Understanding the Importance of Matrices*, 4 x 4 matrices are important in graphics as they allow for easy multiplication of compound transformations. Although I didn't make a big deal of it at the time, these matrices are, in fact, four-dimensional as they have four columns and four rows. Just as we need 4 x 4 matrices to multiply transformation operations, Hamilton found he could use them to find quotients of 3D values. However, the process is a little more complex than how we just created a *w* dimension for coordinates with a 1 or a 0 on the end for *(x, y, z, w)*.

So, where did Hamilton find his fourth dimension? He had to add another number system and he turned to *complex numbers*. If you aren't familiar with complex numbers, then take a look at the explanation here: `https://en.wikipedia.org/wiki/Complex_number`.

In short, complex numbers were devised for solving quadratic equations and to come up with a solution to finding the square root of a negative one. The solution was denoted *i*, thus:

$$i = \sqrt{-1}$$

In this case, *i* is an imaginary number.

With respect to representing 3D coordinates as 4D coordinates, the imaginary axis is added. Complex numbers contain an imaginary part and a real part. The real part is 3D space, as you already understand it. Complex numbers are written with both parts thus:

$$a + bi$$

Representing this on a 2D graph looks like the image in *Figure 16.4*. A point is defined by a horizontal component, which is the real part, and an imaginary component, which is the vertical part. This is not unlike using *x* and *y* to define a coordinate in 2D space once you accept the idea that the real component contains the *x*, *y*, and *z* components:

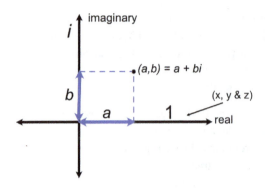

Figure 16.4: Complex numbers

If we rotate a vector of length 1 with values *(a,b)* in this space, the result is as follows:

$$\cos(\theta) + i \times sin(\theta)$$

This is the same as the 2D rotations we examined at the beginning of this chapter and illustrated in *Figure 16.5*. And through this representation, we now have the means to represent a rotation around an arbitrary axis as we did with the Rodrigues-Euler Theorem. The representation of a quaternion is, in fact, an angle and axis stored in four-dimensional space that will allow for compounding operations as the frame of reference remains the same. All rotations occur in the plane *real/i*.

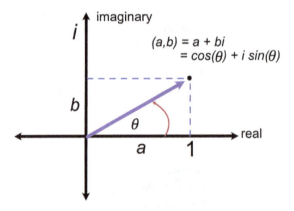

Figure 16.5: Rotating a vector comprising complex numbers

To visualize quaternions is quite difficult as they are in four dimensions, and to show them on a 2D page is even more of a challenge, but bear with me. We live in a 3D world and are familiar with the three-dimensional axes of *x*, *y*, and *z*. With quaternions, we get the addition of a new axis. To see this, imagine that our 3D world is compressed into a flat disk, as shown in *Figure 16.6*. The *x*, *y*, and *z* axes live in this space, and the fourth dimension surrounds the flat disk as a sphere. A quaternion represents a point on the outside of this sphere where the sphere's three dimensions are denoted by three imaginary numbers i, j, and k. A point on the sphere is represented by the coordinate (w, x, y, z) where x is a multiple of i, y is a multiple of j and z is a multiple of k. As an equation, this is as follows:

$$q = w + xi + yj + zk$$

You can clearly see in this equation how x, y, and z are scalar values to the imaginary dimensions.

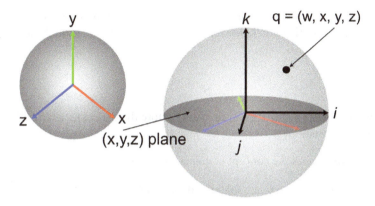

Figure 16.6: Quaternion space

If you aren't already a little mind-bent, the x, y, and z in quaternion coordinates are not the same values for the x, y, and z of the 3D space and yet this is how mathematicians represent them.

Since quaternions are in 4D space, to use them in graphics programming, we need to convert them into values that make sense in 3D. We know that quaternions represent an angle/axis rotation, but how? If we are rotating a point or vector around an axis, r, by an angle of θ then for the quaternion representation $q = (w, x, y, z)$, the value of w is $\cos\left(\frac{\theta}{2}\right)$ and $(x, y, z) = \sin\left(\frac{\theta}{2}\right) \times r$ where r is a vector in

3D space. Here you can see the cosine and sine playing a role in the rotation. The reason why the angle here is divided by two is to align the quaternion representation to that of Euler-Rodrigues' Theorem.

To use a quaternion in our project as a rotation, the quaternion converts into a rotation matrix like this:

$$\begin{bmatrix} 1 - 2y^2 - 2z^2 & 2xy - 2wz & 2xz + 2wy & 0 \\ 2xy + 2wz & 1 - 2x^2 - 2z^2 & 2yz - 2wx & 0 \\ 2xz - 2wy & 2yz + 2wx & 1 - 2x^2 - 2y^2 & 0 \\ 0 & 0 & 0 & 1 \end{bmatrix}$$

More on quaternions

The derivation and proof for quaternions go far beyond the scope of this book; however, the interested reader is encouraged to investigate the topic further through these links:

`https://www.sciencedirect.com/science/article/pii/`
`S0094114X15000415`

```
https://math.stackexchange.com/questions/1385028/concise-
description-of-why-rotation-quaternions-use-half-the-angle
http://www.songho.ca/opengl/gl_quaternion.html
https://www.reedbeta.com/blog/why-quaternions-double-cover/
```

Quaternions, for the novice mathematician, are a very complex and somewhat confusing concept. To really get a feel for them, now that you know the conversion formula, we should start implementing them in our project.

Let's do it...

In this exercise, we will create a `Quaternion` class to store the data associated with a quaternion and enable it to perform rotations.

Follow these steps to implement the concept of quaternions in our project:

1. Make a copy of the `Chapter_15` folder and name it `Chapter_16`.

2. Make a new Python script file and call it `Quaternion.py`. Add the following code to the newly created file:

```python
from __future__ import annotations
import pygame
import math
import numpy as np


class Quaternion:
    def __init__(self, vector=None,
                       axis: pygame.Vector3 = None,
                       angle: float = None):
        if vector is not None:
            self.w = vector[0]
            self.x = vector[1]
            self.y = vector[2]
            self.z = vector[3]
        else:
            axis = axis.normalize()
            sin_angle = \
                math.sin(math.radians(angle/2.0))
            cos_angle = \
```

```
                     math.cos(math.radians(angle/2.0))
            self.w = cos_angle
            self.x = axis.x * sin_angle
            self.y = axis.y * sin_angle
            self.z = axis.z * sin_angle
```

Note the very first line of this code, which refers to `__future__`. For interest's sake, this allows the `Quaternion` class to refer to itself inside the body of the class. That is, it is referencing a `Quaternion` class before the class has been fully described.

The initialization method sets the four values that are used to store a quaternion. The w value is influenced by the cosine operation whereas the x, y, and z values rely on sine. This constructor allows the quaternion properties to be set directly by sending through a vector of four values or by calculating the properties using the axis and angle of rotation.

3. Next, we overload the multiplication operation in the class to define how multiplication between quaternions occurs:

```
    def __mul__(self, other: Quaternion):
        v1 = pygame.Vector3(self.x, self.y, self.z)
        v2 = pygame.Vector3(other.x, other.y, other.z)
        cross = v1.cross(v2)
        dot = v1.dot(v2)
        v3 = cross + (self.w * v2) + (other.w * v1)
        result = Quaternion(vector=(self.w *
                                    other.w - dot,
                                    v3.x, v3.y, v3.z))
        return result
```

As you can see, because quaternions are a complex mathematical construct, they can't be multiplied as easily as numbers in Euclidean space.

4. Last but not least, we can use the quaternion to 4 x 4 matrix conversion to return a matrix we can use to multiply with so it can be integrated with our project's other operations:

```
    def get_matrix(self):
        x2 = self.x + self.x
        y2 = self.y + self.y
        z2 = self.z + self.z
```

```
        xx2 = self.x * x2
        xy2 = self.x * y2
        xz2 = self.x * z2
        yy2 = self.y * y2
        yz2 = self.y * z2
        zz2 = self.z * z2
        wx2 = self.w * x2
        wy2 = self.w * y2
        wz2 = self.w * z2

        return np.matrix([
            [1 - (yy2 + zz2), xy2 + wz2, xz2 - wy2,
             0],
            [ xy2 - wz2, 1 - (xx2 + zz2), yz2 + wx2,
             0],
            [xz2 + wy2, yz2 - wx2, 1 - (xx2 + yy2),
             0],
            [0, 0, 0, 1]
        ])
```

5. To use the new quaternion-based rotation matrix, we must add a new method into the `Transform.py` class like this:

```python
from Quaternion import *
..
def rotate_axis(self, axis: pygame.Vector3, angle,
                    local=True):
    q = Quaternion(axis=axis, angle=angle)
    r_mat = q.get_matrix()
    if local:
        self.MVM = self.MVM @ r_mat
    else:
        self.MVM = r_mat @ self.MVM
```

6. Make a copy of the recent main file called `FlyCamera.py` and rename it `QuaternionTeapot.py`.

7. Edit `QuaternionTeapot.py` to use a quaternion as a rotation in place of our method that used Euler angles like this:

```
teapot = Object("Teapot")
teapot.add_component(Transform())
teapot.add_component(LoadMesh(GL_LINE_LOOP,
                                 "models/teapotSM.obj"))
trans: Transform = teapot.get_component(Transform)

trans.rotate_axis(pygame.Vector3(0, 1, 0), 90)
trans.update_position(pygame.Vector3(0, -2, -3))
```

8. You can now run `QuaternionTeapot.py` and see the teapot rotated 90 degrees around the *y* axis. The result will be the same as the previous code, as illustrated in *Figure 16.7*, in which I used the *S* key to move the camera away from the teapot to fit it in the window:

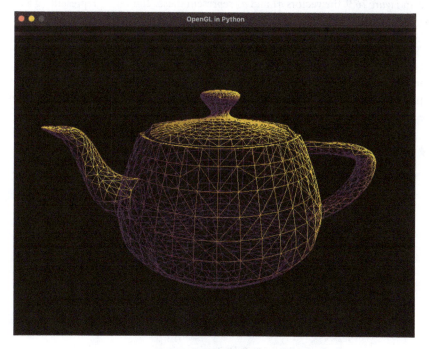

Figure 16.7: Rotating the teapot with a quaternion

In this exercise, we have validated rotating an object with a single quaternion by recreating the *y* axis rotation we achieved in *Chapter 15, Navigating the View Space*. When introducing a new mathematical method, to ensure it behaves as you believe it should, always find another way to achieve the same result if possible.

This section has been a whirlwind look at quaternions. As you will appreciate, they are a highly complex mathematical construct. Even if you don't fully understand all the mathematics involved, it has been my intention that you will at least appreciate what they can do and how they can be implemented in code. However, we didn't just add quaternions in to make single rotations more complex; their power in graphics is to remove any issues related to the gimbal lock inherent with Euler angles. We will now discuss this topic and modify our project appropriately.

Working with unit quaternions

As discussed, quaternions remove the limitations involved in compounding Euler-angle rotations. In this section, we will concentrate on reprogramming the camera in our project to pitch and roll with quaternions.

Before quaternions are multiplied, we must ensure they are **unit quaternions**. That means they will have a length of 1. If we go back to thinking of quaternion spacing being a sphere encompassing Euclidean space, then a quaternion represents a vector from the origin of both spaces to the surface of the sphere. In *Figure 16.9*, the vectors *q1* and *q2* represent these. The vector representing a quaternion is four-dimensional, with the coordinates storing the angle and axis as $q = (w, x, y, z)$.

Of course, you must remember we can't see these four dimensions in our 2D/3D sphere diagram, but they are there. In *Figure 16.8*, *q3* represents an invalid quaternion in that it extends beyond the surface of the sphere. However, the rotation that it is meant to represent can be found by normalizing its vector representation back to a length of 1, shown by vector *q4*. This unit length vector now points nicely to the surface of the sphere, or quaternion space, instead of beyond it:

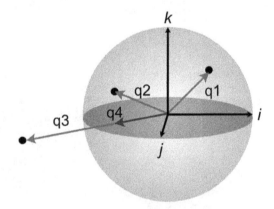

Figure 16.8: Unit quaternions

The next question you should be asking is, *"how do I normalize a quaternion?"* It's a matter of finding the length of the quaternion and dividing each component by that length in the same way we normalized vectors in *Chapter 9, Practicing Vector Essentials*. To find the length of a quaternion, we can use Pythagoras' Theorem, where the length of a quaternion is calculated with the following:

$$length = \sqrt{w^2 + x^2 + y^2 + z^2}$$

The normal form of the quaternion is then:

$$q = (\frac{w}{length}, \frac{x}{length}, \frac{y}{length}, \frac{z}{length})$$

Let's see if you can now use this knowledge to create a method to produce a normal vector.

Your turn...

Exercise A:

Using your understanding of normalizing vectors, write down on a piece of paper pseudocode to produce a normalized quaternion.

In this section, we have discussed the importance of unit quaternions. But why aren't they always in unit form? Let's take a look.

Understanding the purpose of normalization

If you are wondering why we need to normalize a quaternion in the first place, it's because the values used to create it may produce a quaternion with a length longer than 1, especially if it's code that you are adding yourself.

Take, for example, the 3D process of moving an object along a vector at a constant speed. In *Chapter 10, Getting Acquainted with Lines, Rays, and Normals*, we moved an object along a line segment at a constant speed. To achieve a constant speed, we needed to take into consideration the time between frames so we could factor in any changes. The code we created moved an object in equal steps from one end of a line segment to the other. Taking this same idea, we can write code to move an object in 3D along a vector at a constant speed. The essential parts of this script would look something like this:

```
dt = 0
direction = (0, 0, 0.1)
while not done:
    new_position = old_position + (direction * dt)
    dt = clock.tick(fps)
```

In this case, the object doesn't have an ending location; it just moves along the `direction` vector multiplied by the time between frames each frame. Now, if `direction` becomes a calculated value, say we want to move the object between its current location toward another object, the length of `direction` will become crucial. Consider the tanks in *Figure 16.9*. They are to move toward the palm tree. The direction of the red tank to the tree is calculated with the following:

```
v = tree_position - red_tank_position
```

This situation is visualized in *Figure 16.9*:

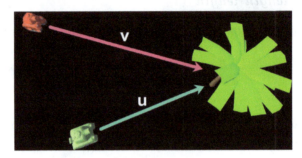

Figure 16.9: Two objects moving toward a destination

Likewise, the vector u is calculated to be as follows:

```
u = tree_position - green_tank_position
```

The distance the red tank is from the tree will be the length of the vector v, and the distance the green tank is from the tree will be the length of the vector u. For the sake of this example, let's say that the length of v is 10, and the length of u is 7.

If we move each tank along its associated vector in one frame, the red tank will move 10 units per frame and the green tank, 7 units per frame. We can agree that makes the red tank faster as it covers more distance in the same amount of time. If, however, we want both these tanks to move at the same speed, we can find the unit vectors in the direction of travel. As both unit vectors, no matter which way they are facing, have a length of 1, when we add 1 to both tanks in each frame, they will be traveling at the same speed. Even if we multiply their unit direction vectors by the time between frames, they will still move at the same rate. A partial listing of the logic of this code looks like this:

```
dt = 0
red_direction = tree_position - red_tank_position
green_direction = tree_position - green_tank_position
red_normalised = red_direction/red_direction.magnitude
green_normalised =
    green_direction/green_direction.magnitude
```

```
while not done:
  red_tank_position = red_tank_position + (red_normalised *
                                          dt)
  green_tank_position = green_tank_position +
                        (green_normalised * dt)
  dt = clock.tick(fps)
```

This example shows where unit vectors are useful and how they are calculated in code. The same issue can also occur when human input is required. Let's say you want the tanks to move in a 45-degree direction on the x/z plane, as shown in *Figure 16.10*.

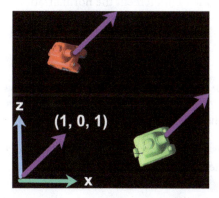

Figure 16.10: Specifying a movement direction

A 45-degree vector on the x/z plane can be denoted (1, 0, 1). This makes sense as it has equal parts of x as it does of z, hence cutting the 90-degree angle between the *x* axis and the *z* axis in half. Similarly, the vector (0.5, 0, 0.5) and (3, 0, 3) are also 45-degree vectors pointing in the same direction. Though, which of all these is a unit vector? The answer is none of them. If you normalize each of these vectors, you will find the unit vector in the same direction:

$$length\ of\ (1,0,1) = \sqrt{1^2 + 0^2 + 1^2} = \sqrt{2}$$

$$normalize\ (1,0,1) = \frac{(1,0,1)}{\sqrt{2}} = (0.707, 0, 0.707)$$

$$length\ of\ (0.5,0,0.5) = \sqrt{0.5^2 + 0^2 + 0.5^2} = \sqrt{0.5}$$

$$normalize\ (0.5,0,0.5) = \frac{(0.5,0,0.5)}{\sqrt{0.5}} = (0.707, 0, 0.707)$$

$$length\ of\ (3,0,3) = \sqrt{3^2 + 0^2 + 3^2} = \sqrt{18}$$

$$normalize\ (3,0,3) = \frac{(3,0,3)}{\sqrt{18}} = (0.707, 0, 0.707)$$

Notice how all the calculations result in the same vector? That's because all the original vectors pointed in the same direction.

So, whether you find it easier to manually set a vector using your intuition, such as using (1, 0, 1) instead of (0.707, 0, 0.707), or a distance calculation happens in the code, and then a unit vector needs to be found, being able to normalize a vector is a useful operation.

The same goes for quaternions. Possibly even more so, as it's going to be more intuitive to say you want to rotate an object around the axis (1, 0, 1) with an angle of 45 than being concerned if the resulting quaternion has a length of 1. Therefore, being able to normalize vectors and quaternions is key for consistency in mathematics.

This brings us back to the matter of being able to normalize a quaternion, which we achieved in the previous section. Now it's time to implement the normalization in our project and use it for compounding rotations.

Let's do it...

In this exercise, we will write up the code for normalizing a quaternion and then use it to ensure all quaternions that are multiplied together are a length of 1. Having achieved this, we will begin testing the multiplication overload in the Quaternion class:

1. To the Quaternion class in Quaternion.py, add a normalize method and use it when a quaternion is constructed, as follows:

```
import sys

..

class Quaternion:
def __init__(self, vector=None, axis: pygame.Vector3 =
    None, angle: float = None):

    if vector is not None:

        ..

    else:
        axis = axis.normalize()

        ..

        self.z = axis.z * sin_angle
    self.normalise()


..
def normalise(self):
```

```
                    length = math.sqrt(self.w * self.w +
                                        self.x * self.x +
                                        self.y * self.y +
                                        self.z * self.z)
                    if length > sys.float_info.epsilon:
                        self.w /= length
                        self.x /= length
                        self.y /= length
                        self.z /= length
```

In this method, the length of the quaternion is found using Pythagoras' Theorem. This length is then compared to `sys.float_info.epsilon`. This system variable holds a very small floating-point number that is close to 0. Sometimes when working with floats, a value that would equate to 0 on paper might be stored in the computer fractionally larger than 0. As the length of an essentially 0-length quaternion might be calculated as something close to zero before we divide by the length to get the normalized quaternion, we test that the quaternion has a decent length using epsilon instead of 0.

2. To compare the results from using quaternions to those of Euler angles, we will now add two teapots into `QuaternionTeapot.py` and rotate one with Euler angles and one with quaternions and then compare the results.

3. Before you do this, you'll need to add a new method to the `Transform` class in `Transform.py`, as follows:

```
def rotate_quaternion(self, quaternion: Quaternion,
                      local=True):
    r_mat = quaternion.get_matrix()
    if local:
        self.MVM = self.MVM @ r_mat
    else:
        self.MVM = r_mat @ self.MVM
```

4. Modify `QuaternionTeapot.py`, as follows:

```
..
objects_3D = []
objects_2D = []
```

```
#onleft
euler_teapot = Object("Teapot")
euler_teapot.add_component(Transform())
euler_teapot.add_component(LoadMesh(GL_LINE_LOOP,
                          "models/teapotSM.obj"))
euler_trans: Transform =
    euler_teapot.get_component(Transform)
euler_trans.rotate_x(45)
euler_trans.rotate_y(20)
euler_trans.update_position(pygame.Vector3(-3, 0,
                                          -10))
print("Euler Rot")
print(euler_trans.get_rotation())


#onright
quat_teapot = Object("Teapot")
quat_teapot.add_component(Transform())
quat_teapot.add_component(LoadMesh(GL_LINE_LOOP,
                          "models/teapotSM.obj"))
quat_trans: Transform =
    quat_teapot.get_component(Transform)
q = Quaternion(axis=pygame.Vector3(1, 0, 0), angle=45)
p = Quaternion(axis=pygame.Vector3(0, 1, 0), angle=20)
t = p * q
quat_trans.rotate_quaternion(t)
quat_trans.update_position(pygame.Vector3(3, 0, -10))
print("Quat Rot")
print(quat_trans.get_rotation())


camera = Camera(60, (screen_width / screen_height),
              0.1, 1000.0)

camera2D = Camera2D(gui_dimensions[0],
          gui_dimensions[1], gui_dimensions[3],
          gui_dimensions[2])
```

```
objects_3D.append(euler_teapot)
objects_3D.append(quat_teapot)

clock = pygame.time.Clock()


..
```

This new code adds two teapots: one called `euler_teapot` and one called `quat_teapot`. Now, `euler_teapot` is positioned on the left of the screen, and `quat_teapot` is on the right. Both teapots are first rotated by 45 degrees around the *x* axis and then 20 degrees about the *y* axis.

Note that the multiplication for the quaternions occurs in reverse order, such as we used to multiply the transformation matrices in the line `t = p * q`.

After the rotations have been applied to each teapot, the rotation matrices are printed out for you to compare the results, as sometimes visual cues don't necessarily give us enough accuracy.

5. Run the script to see the two teapots, as illustrated in *Figure 16.11*. Have they been rotated in the same manner? It is difficult to tell from a perspective view; however, they are facing in similar directions. That's a good start. However, if you really want to know if the quaternion equations match those of the Euler calculations and vice versa, the only true way to check is to print out the rotation matrices for each, which will be displayed in the console.

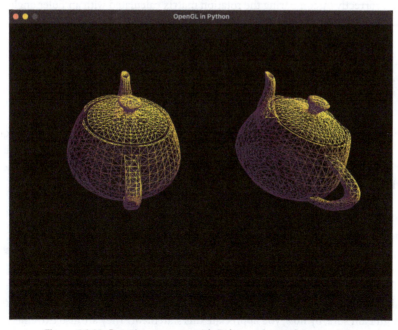

Figure 16.11: Rotating teapots with Euler angles and quaternions

In this case, my console is displaying the following:

```
Euler Rot
[[ 0.93969262    0.           -0.34202014    0.          ]
 [ 0.24184476    0.70710678    0.66446302    0.          ]
 [ 0.24184476   -0.70710678    0.66446302    0.          ]
 [ 0.           0.            0.            1.          ]]
Quat Rot
[[ 9.39692621e-01   1.38777878e-17  -3.42020143e-01   0.0000e+00]
 [ 2.41844763e-01   7.07106781e-01   6.64463024e-01   0.0000e+00]
 [ 2.41844763e-01  -7.07106781e-01   6.64463024e-01   0.0000e+00]
 [ 0.00000000e+00   0.00000000e+00   0.00000e+00   1.0000e+00]]
```

Yours might display differently if you are on Windows, and you might even have slightly different values to mine. While at first glance, the values look different, they are almost the same. For example, my quaternion matrix has $1.38777878e-17$, whereas the Euler matrix has 0. I'm not at all concerned as 1.3×10^{-17} is practically zero, and because we are working with floating-point numbers and performing differing calculations, there are bound to be slight errors. However, these are small enough for us to conclude the quaternion calculations are working as intended.

In this exercise, we have integrated compound quaternion rotations into our project and confirmed they are working correctly by comparing the rotation matrix calculated with an Euler angle calculated based on the same axes and rotations. Now it's time for you to test your understanding by replacing the Euler movements for the camera with quaternions.

Your turn...

Exercise B:

Reprogram the `rotate_with_mouse()` method of the `Camera` class to use a compound quaternion rotation that multiplies the x rotation with the y rotation and then applies it to the camera. The axis of rotation for the x will be the right axis, and for the y, it will be the camera's up axis. These can be found in the view matrix, as follows:

```
right = pygame.Vector3(1,0,0)
up = pygame.Vector3(0,1,0)
```

Summary

In case you missed it the first time, let me say it again: *quaternions are an advanced mathematical construct*. Though I am sure, by now, you appreciate this statement. They are also extremely powerful, and this chapter has but scratched the surface of all the applications for which they can be applied.

Hamilton wasn't even thinking of 3D graphics rotations when he defined them, but thankfully for us, they exist and remove the inherent issue of compounding Euler angle rotations.

If you've reached the end of this chapter and still don't feel comfortable employing quaternion mathematics, you won't be alone. In fact, I hesitated to include this chapter as a full comprehension of quaternions requires background knowledge in complex numbers, pure mathematics, and division algebra that we don't have the scope in this book to include. And if you don't feel comfortable yet working them, then the simple solution is, don't. Euler angles will achieve most things you want to do when it comes to rotation. Having said this, as you continue your learning journey in graphics and games programming, you won't be able to avoid them.

This chapter concludes our exploration of transformations. In the next chapter, we will begin investigating the process of rendering and production of visual effects for graphics.

Answers

Exercise A:

```
normalize(Quaternion q)
  length = sqrt( pow(w,2) + pow(x,2) + pow(y,2) + pow(z,2))
  q.w /= length
  q.x /= length
  q.y /= length
  q.z /= length
```

Exercise B:

```
def rotate_with_mouse(self, yaw, pitch):
    right = pygame.Vector3(1, 0, 0)
    up = pygame.Vector3(1, 0, 0)
    y_rot = Quaternion(axis=up, angle=self.mouse_invert * yaw *
                       self.mouse_sensitivityY)
    x_rot = Quaternion(axis=right,
                       angle=self.mouse_invert * pitch *
                       self.mouse_sensitivityX)
    m_rot = x_rot * y_rot
    self.transform.rotate_quaternion(m_rot)
```

In the preceding code, the camera is rotated around its own up axis instead of the world axis. When you test this out, the movement might seem quite strange if you are used to navigating using the previous version of the method and rotating around the world up. If you aren't quite sure it's working correctly, use `x_rot` and `y_rot` one at a time to rotate and test the individual movements. For example, run it once with:

```
self.transform.rotate_quaternion(x_rot)
```

Then run it again with:

```
self.transform.rotate_quaternion(y_rot)
```

Remember, depending on how you like the mouse to rotate the world, you can change the value of `self.mouse_invert`.

Part 4 – Essential Rendering Techniques

This part covers the visual representation of 3D objects and introduces vertex and fragment shaders. Through the use of shaders, you will discover how to optimize and speed up graphics rendering by employing the power of the graphics processing unit. The contents will cover the practical conversion of the OpenGL program written throughout to use shader code. In addition, shading models for lighting will be applied before the book wraps up by implementing the latest shader technique of physically-based rendering.

In this part, we cover the following chapters:

- *Chapter 17, Vertex and Fragment Shading*
- *Chapter 18, Customizing the Render Pipeline*
- *Chapter 19, Rendering Visual Realism Like a Pro*

17
Vertex and Fragment Shading

The rendering of the models we've achieved thus far has used OpenGL's deprecated API calls along with mathematics calculated on the CPU by Python to draw, texture, and light images on the screen. If you have tried to render models with many vertices using the current project, you'll have noticed how slow the methods become as the vertex count increases. Even the original teapot model from *Chapter 8, Reviewing Our Knowledge of Triangles*, starts to slow down the application.

To speed up rendering, a graphics card (GPU) can be used to process vertices and pixels in parallel. To move the logic and algorithms we've written thus far onto the GPU, we must first learn how to write shader programs that are compiled and executed on the GPU.

To this end, in this chapter, we will cover the following topics:

- Understanding shaders
- Transferring processing from the CPU to the GPU
- Processing pixel by pixel

As we refactor the code base of our project to work with shaders, you'll see how the mathematics we've covered thus far remains relevant. You will learn how mathematical operations can be performed on the GPU to convert a vertex in model space into screen space. By moving more mathematical operations out of Python and into shader code, you can make your rendering process highly efficient as they become operations processed in parallel rather than procedural operations on the CPU. This knowledge will help you become a more skilled graphics programmer capable of generating highly optimized rendering programs.

We will begin the chapter by examining the render pipeline and discussing where shader code can be used to customize the process.

Technical requirements

The solution files containing the code can be found on GitHub at `https://github.com/PacktPublishing/Mathematics-for-Game-Programming-and-Computer-Graphics/tree/main/Chapter17` in the `Chapter17` folder.

Understanding shaders

In graphics, a **shader** is a special program used to determine how a virtual object is drawn on the screen. It calculates the color of each pixel based on the geometry of an object, the color or texture of the object, and the light falling on the object's surface. Shaders are written in a **shading language**. There are three general types of shaders—**fragment**, **vertex**, and **geometry**. These shaders allow graphics programmers to intercept the render pipeline shown in *Figure 17.1*:

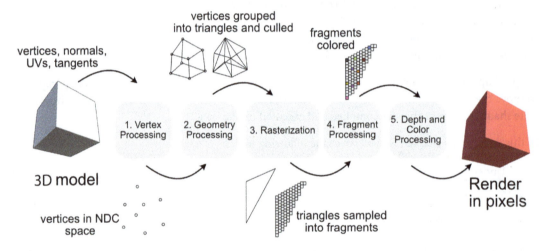

Figure 17.1: The render pipeline

The process of rasterization is how 3D virtual objects get drawn onto the screen as pixels. It starts out as a 3D mesh or model with vertices, normals, UVs, and tangents. The vertex processor (**1** in *Figure 17.1*) computes the normalized coordinate-space positions of the vertices and associated values, taking them from model coordinates and converting them into screen coordinates. The mathematics for this process was revealed in *Chapter 4, Graphics and Game Engine Components*. After leaving the vertex processor, the vertices are grouped together into primitive triangles during geometry processing (**2**), and they are clipped based on their position in the world and where they are in the camera view. The surviving triangles are rasterized (**3**). This process produces a series of fragments. Fragments are like pixels, in that they represent a point on the screen. However, fragments are only *potential pixels* as they may be further culled when the fragment of one object is obscured by another. The fragment representation of a triangle is transformed during fragment processing (**4**) into colored fragments, which are then compared against the depth and color of other fragments (**5**) in the frame buffer to produce a final rendered pixel.

The programmable parts of this pipeline can be customized using shader code to not only render a model but to also add special effects. During (**1**), a **vertex shader** transforms each vertex's 3D position in model space to the 2D coordinate of its associated fragment in screen space and possible pixel, as shown in *Figure 17.2*:

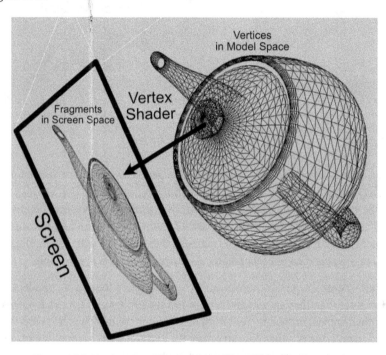

Figure 17.2: Vertices transformed from 3D to 2D by a vertex shader

With this information, the depth of that fragment is also stored to allow for the culling of fragments to occur later. Vertex shaders can change a vertex's position, color, and UV coordinates. No additional vertices can be added by a vertex shader. The output from (**1**) is passed to (**2**), where a **geometry shader** can be used to manipulate the vertices and even add more vertices. Using a geometry shader is a relatively new addition to the render pipeline and is not supported by all graphics cards or APIs. At this point, if a shader is not custom-defined by a programmer, this step acts in a default manner to construct triangles. After the rasterization process (**3**) has constructed fragments, they are able to be customized by a **fragment shader** that provides color and shadowing, as shown in *Figure 17.3*:

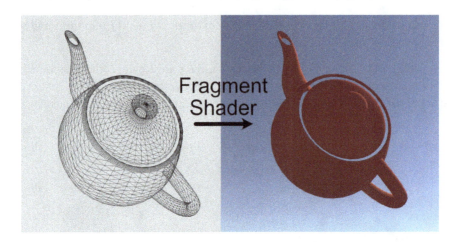

Figure 17.3: Fragment shader

These shaders run in parallel on the GPU. The vertex and geometry shader run once for each vertex in the model, processing each one separately. The key thing to note here is that because the shader is running for a single vertex inside the shader, it does not have access to the vertices for any comparison processing. However, because all shaders run in parallel, many vertices can be processed at the same time. The fragment shader runs once for each fragment that is passed to it. Again, the runs happen in parallel, and accessing the data of other fragments during the run is not possible.

When you first start writing shaders, it can take a little while to grasp the way the shaders are working. Often, in other types of programming, whether it be procedural or object-oriented, it is always possible to access data of other similar constructs. For example, given a mesh made up of several vertices, these vertices are stored in an array or other data structure. In Python, if the vertices are in an array, then it's easy to compare the vertex in array position 0 with another vertex in array position 10. This is not possible in shader code, although some tricks can be performed to pass extra data and information to a running shader.

Expanding shader functionality

In addition to the traditional shaders of vertex, geometry, and fragment, as technology progresses, more shader functionality becomes available. To date, there are other types of shaders that are available, including the following:

Compute shaders (`https://learnopengl.com/Guest-Articles/2022/Compute-Shaders/Introduction`)

Tessellation Evaluation Shaders (`https://www.khronos.org/opengl/wiki/Tessellation_Evaluation_Shader`)

Tessellation Control Shaders (`https://www.khronos.org/opengl/wiki/Tessellation_Control_Shader`)

On completion of this chapter, once you've gained an understanding of how shaders work and can be integrated into your graphics programs, you are encouraged to do independent research on including these shaders in your own projects.

In this section, we've taken a close look at the rasterization process and where custom programs can be inserted to modify the resulting image. In the next section, you will get a taste of creating these shaders to begin controlling the drawing processes occurring on the GPU.

Transferring processing from the CPU to the GPU

OpenGL pre-dates modern graphics cards as we know them now. It was originally built to work with render farms, which were mostly superseded by modern-day graphics cards. As such, it underwent an API redesign in 2008 with the release of version 3.0, which focused on GPU programming.

You might now be wondering whether what you've learned thus far in this book is suddenly out of date! The answer is *no*. The mathematics is always relevant—it never changes. Up until now, you've received a gentle introduction to graphics programming. From here on, it gets more challenging.

If you think about it, the transforms that we've performed on an object move the object as a whole, but the equations for the transforms move each vertex individually. It's just the nature of affine transformations. As discussed in *Chapter 12*, *Mastering Affine Transformations*, this means that moving an object is the same as transforming each of its vertices.

Before we get started on the mathematics, we should modify the project we've been working on to deal with this new way of processing graphics. There's a bit of tweaking that needs to be done, so bear with me.

Let's do it...

Follow the next steps:

1. Make a copy of the `Chapter_16` folder, and name it `Chapter_17`.

2. For OpenGL to process shader code, it must process and compile it for the GPU. To do this, we need to write a dedicated function. Open `Utils.py`. It will only have the `map_value()` function in it. Add the following code:

```
from OpenGL.GL import *
import numpy as np

def map_value(current_min, current_max, new_min,
              new_max, value):
 . .
```

```python
def compile_shader(shader_type, shader_source):
    shader_id = glCreateShader(shader_type)
    glShaderSource(shader_id, shader_source)
    glCompileShader(shader_id)
    compile_success = glGetShaderiv(shader_id,
                                    GL_COMPILE_STATUS)
    if not compile_success:
        error_message = glGetShaderInfoLog(shader_id)
        glDeleteShader(shader_id)
        error_message = "\n" +
                        error_message.decode("utf-8")
        raise Exception(error_message)
    return shader_id
```

This first method compiles the shader code. It contains four OpenGL methods that are critical to this task. First, glCreateShader() allocates memory to store a new shader that is referenced by shader_id. glShaderSource() takes the shader script we will write in separate text files in the *Processing pixel by pixel* section, and places it in the newly created shader. glCompileShader() then compiles the shader into machine code that can be understood by the GPU. The remainder of the method checks on the status of the compiler and reports any errors you might have made in the shader code.

Next, we need to program a method to turn the compiled shaders into programs that we can access in code as needed. This method is shown here:

```python
def create_program(vertex_shader_code,
                   fragment_shader_code):
    vertex_shader_id =
        compile_shader(GL_VERTEX_SHADER,
                       vertex_shader_code)
    fragment_shader_id =
        compile_shader(GL_FRAGMENT_SHADER,
                       fragment_shader_code)
    program_id = glCreateProgram()
    glAttachShader(program_id, vertex_shader_id)
    glAttachShader(program_id, fragment_shader_id)
```

```
glLinkProgram(program_id)
link_success = glGetProgramiv(program_id,
                                GL_LINK_STATUS)
if not link_success:
    info = glGetProgramInfoLog(program_id)
    raise RuntimeError(info)
glDeleteShader(vertex_shader_id)
glDeleteShader(fragment_shader_id)
return program_id
```

Here, in `create_program()`, the text files containing our shader code are passed through to the `compile_shader()` function. We will be writing both vertex and fragment shaders. `glCreateProgram()` creates a new program object that will link our program to the shaders. Because the shaders are separately compiled items, you must consider them programs outside of the Python code. Therefore, we need to link to them. As you can see in the preceding code snippet, `program_id` has both the vertex shader and fragment shader attached. Then, `program_id` is used to point our Python program to the executable versions of the vertex and fragment shaders. Following this setup, the status of creating the program can be verified and errors reported. Finally, `glDeleteShader()` is called twice to remove the shader source from memory. Once it is contained in the GPU, there's no need to clog up memory with the shader source.

3. Shaders are linked to models by the use of **materials**. A material becomes a component of a mesh, just like a transform. It specifies how the mesh is to be rendered. This means that you can select which shader to use on a model as well as change it. Different models can have different shaders. Create a new Python script called `Material.py` and add the following code to it:

```
from Utils import *


class Material:
    def __init__(self, vertex_shader,
                    fragment_shader):
        self.program_id = create_program(
                    open(vertex_shader).read(),
                    open(fragment_shader).read())


    def use(self):
        glUseProgram(self.program_id)
```

The material will take the vertex and fragment code that we create, and run it through the shader creation methods we just added to `Utils.py`. The `glUseProgram()` function sets the current rendering state to use the associated shader code to process the drawing of the model.

4. To use shaders in our project, the main script needs to be changed. Create a new Python script called `ShaderTeapot.py`. All of the following code will be new to this file, though I have presented the shader-specific code in bold for you to take extra note of as you type it out:

```python
from Object import *
from pygame.locals import *
from Camera import *
from LoadMesh import *
from Material import *
from Settings import *

pygame.init()
pygame.display.gl_set_attribute(
    pygame.GL_MULTISAMPLEBUFFERS, 1)
pygame.display.gl_set_attribute(
    pygame.GL_MULTISAMPLESAMPLES, 4)
pygame.display.gl_set_attribute(
    pygame.GL_CONTEXT_PROFILE_MASK,
    pygame.GL_CONTEXT_PROFILE_CORE)
pygame.display.gl_set_attribute(pygame.GL_DEPTH_SIZE,
                                32)
```

The `gl_set_attribute()` calls set up the pygame environment to work with the OpenGL shaders. The first two, which include `MULTISAMPLE`, set up the environment to help reduce jaggedness or aliasing that you find when drawing lines on a screen. You can read more about this here: `https://learnopengl.com/Advanced-OpenGL/Anti-Aliasing`.

The `CONTEXT_PROFILE` settings tell pygame to use the current OpenGL contents, which in short accesses the shader functionality in versions of OpenGL 3.0 and after. When this is set, you'll find that any old, depreciated OpenGL calls in your script will cause errors, and we will be deleting them soon. For more information on OpenGL contexts, see `https://www.khronos.org/opengl/wiki/OpenGL_Context#OpenGL_3.2_and_Profiles`. The last attribute set is `GL_DEPTH_SIZE`. In this case, we are setting it to use `32`-bit floating values. *If this setting causes issues on a Windows machine, you can leave it out.* However, on macOS, if it is not specified, the depth testing performed by OpenGL is substandard. We rely on depth testing to ensure pixels get drawn in the correct order.

The remainder of this code is like that used in `QuaternionTeapot.py` from *Chapter 16, Rotating with Quaternions*, with the exception that a material has been created and added as a component to the teapot, as shown here:

```python
screen_width = math.fabs(window_dimensions[1] -
                            window_dimensions[0])
screen_height = math.fabs(window_dimensions[3] -
                            window_dimensions[2])
pygame.display.set_caption('OpenGL in Python')
screen = pygame.display.set_mode((screen_width,
                                    screen_height),
                                    DOUBLEBUF | OPENGL)
done = False
white = pygame.Color(255, 255, 255)
glDisable(GL_CULL_FACE)
glEnable(GL_DEPTH_TEST)
glEnable(GL_BLEND)
glBlendFunc(GL_SRC_ALPHA, GL_ONE_MINUS_SRC_ALPHA)
objects_3d = []
camera = Camera(60, (screen_width / screen_height),
                0.01, 10000.0)


quat_teapot = Object("Teapot")
quat_teapot.add_component(Transform())
mat = Material("shaders/vertexcolvert.vs",
                "shaders/vertexcolfrag.vs")
quat_teapot.add_component(LoadMesh(quat_teapot.vao_ref
    , mat, GL_LINE_LOOP, "models/teapot.obj"))
quat_teapot.add_component(mat)
quat_trans: Transform =
    quat_teapot.get_component(Transform)
quat_trans.update_position(pygame.Vector3(3, 0, 10))


objects_3d.append(quat_teapot)


clock = pygame.time.Clock()
```

```
fps = 30

pygame.event.set_grab(True)
pygame.mouse.set_visible(False)
while not done:
    events = pygame.event.get()
    for event in events:
        if event.type == pygame.QUIT:
            done = True
        if event.type == KEYDOWN:
            if event.key == K_ESCAPE:
                pygame.mouse.set_visible(True)
                pygame.event.set_grab(False)
            if event.key == K_SPACE:
                pygame.mouse.set_visible(False)
                pygame.event.set_grab(True)

    glClear(GL_COLOR_BUFFER_BIT | GL_DEPTH_BUFFER_BIT)
    camera.update()
    for o in objects_3d:
        o.update(camera, events)

    pygame.display.flip()
    dt = clock.tick(fps)
pygame.quit()
```

There's also a modification to the `LoadMesh()` method call with the addition of `quat_teapot.vao_ref`, which we will add shortly.

5. Next, we create two helper classes to allocate and structure memory to hold the data to be used by the shaders. The shader code we will write will replace all the OpenGL drawing code we've written to date. In some ways, the shader code will also simplify the OpenGL drawing code, but to work, the shaders still need to know about vertices, colors, projection and view matrices, and other data. The first class we create will be in a script called `GraphicsData.py`. Create this file and add the following code to it:

```
from OpenGL.GL import *
import numpy as np
```

```
class GraphicsData():
    def __init__(self, data_type, data):
        self.data_type = data_type
        self.data = data
        self.buffer_ref = glGenBuffers(1)
        self.load()


    def load(self):
        data = np.array(self.data, np.float32)
        glBindBuffer(GL_ARRAY_BUFFER, self.buffer_ref)
        glBufferData(GL_ARRAY_BUFFER, data.ravel(),
                     GL_STATIC_DRAW)


    def create_variable(self, program_id,
                            variable_name):
        variable_id = glGetAttribLocation(program_id,
            variable_name)
        glBindBuffer(GL_ARRAY_BUFFER, self.buffer_ref)
        if self.data_type == "vec3":
            glVertexAttribPointer(variable_id, 3,
                                  GL_FLOAT,
                                  False, 0, None)
        elif self.data_type == "vec2":
            glVertexAttribPointer(variable_id, 2,
                                  GL_FLOAT,
                                  False, 0, None)


        glEnableVertexAttribArray(variable_id)
```

This code creates a new variable to store data passed to it, format it, and load it into an OpenGL buffer. The data is stored in GL_ARRAY_BUFFER. In the load() method, glBindBuffer() causes buffer_ref (short for *buffer reference*) to point at the memory location where the data sits. The original data is placed in an np.array, which is then placed into its allocated buffer using glBufferData().data.ravel() flattens the array into a continuous string of floating-point values. This means any row and column formatting is lost.

Once the data is in a long string of values, if the format of these values is not known, it is impossible to know where one number ends and another begins. Consider the following example of the data being originally in this format:

$$\begin{bmatrix} (1,2,4) & (3,5,6) \\ (9,2,6) & (1,4,5) \end{bmatrix}$$

When flattened, it would look like this:

124356926145

Now, consider that if instead of 3D values, the data looked like this:

$$\begin{bmatrix} (1,2,) & (5,6) \\ (9,6) & (1,4) \end{bmatrix}$$

The flattened version would look like this:

12569614

In both flattened versions, it's impossible to know whether the first value is 1 or 12 unless you know how big each original number was and which format it was in.

Therefore, in the `create_variable()` method, `glVertexAttribPointer()` provides a means of letting OpenGL know the size and dimensionality of each value in the data array. The first use of it creates an array of floats, where each set of values comes in threes. This is how 3D vectors are stored in a flattened array and how OpenGL knows where each of the x, y, and z values are positioned. The same happens for 2D values in the second use of `glVertexAttribPointer()`. For more details, check out `https://registry.khronos.org/OpenGL-Refpages/gl4/html/glVertexAttribPointer.xhtml`.

The data arrays created in this class are fixed and don't change over the course of the program. For example, the vertices of the model aren't going to change, and therefore, they can be stored in this way. However, it is possible to have variable values in shaders, and we will now create a separate class to deal with them.

6. Variable values that can change and be passed to a shader are called **uniforms**. Create a new Python script called `Uniform.py` and add the following code to it:

```python
from OpenGL.GL import *


class Uniform():
    def __init__(self, data_type, data):
        self.data_type = data_type
        self.data = data
```

```
        self.variable_id = None

    def find_variable(self, program_id,
                    variable_name):
        self.variable_id =
            glGetUniformLocation(program_id,
                                variable_name)

    def load(self):
        if self.data_type == "vec3":
            glUniform3f(self.variable_id,
                        self.data[0],
                        self.data[1], self.data[2])
        elif self.data_type == "mat4":
            glUniformMatrix4fv(self.variable_id, 1,
                            GL_TRUE,
                            self.data)
        elif self.data_type == "sampler2D":
            texture_obj, texture_unit = self.data
            glActiveTexture(GL_TEXTURE0 +
                        texture_unit)
            glBindTexture(GL_TEXTURE_2D, texture_obj)
            glUniform1i(self.variable_id,
                        texture_unit)
```

You'll notice that the code for the uniform is very similar for the code in GraphicsData.py; however, instead of having glVertexAttribPointer store data in any format by specifying the datatype and sizes, uniforms have special calls based on the datatype. glUniform3f() will create a variable that stores three floating-point values, glUniformMatrix4fv() stores a 4x4 matrix of floats, and glUniform1i() stores long integer values that can represent data in a texture.

The difference between the fixed values stored by glVertexAttribPoint and glUniform is that the former is a long array of multiple vertex values, such as a list of all the vertices that represent a model, whereas a uniform value is just one item—for example, a matrix or a single texture.

For more details on all `glUniform()` methods, see `https://registry.khronos.org/OpenGL-Refpages/gl4/html/glUniform.xhtml`.

> **OpenGL shader programming**
>
> At this point, most of the refactoring of our code base that is required for moving forward with vertex and fragment shaders has been done. I realize there are a lot of new ideas presented here and would encourage you to seek further elucidation with the following resources (if desired) before we start the shader programming:
>
> `https://learnopengl.com/Getting-started/Shaders`
>
> `https://www.khronos.org/opengl/wiki/Shader`
>
> `https://www.lighthouse3d.com/tutorials/glsl-tutorial/`
>
> However, if you would like to obtain a copy of the code base up until this point before we continue, you can download a copy from GitHub at `https://github.com/PacktPublishing/Mathematics-for-Game-Programming-and-Computer-Graphics/tree/d2629ef63350c8ae0cba3dfba6f3dcb4a4bf6f62/Chapter%2017/starter`.

In this section, we've spent a lot of time refactoring the code base. This has been necessary to move on to the next step of introducing shader coding. While many books might concentrate on shader coding from the get-go, a lot of the mathematics presented herein is easier to grasp through its introduction step by step, which OpenGL up until this point has allowed us to do.

In the next section, we take a big step forward into rendering using the GPU, which means learning to program vertex by vertex and pixel by pixel and shifting a lot of the mathematics into shaders.

Processing pixel by pixel

Shaders are run on the GPU in parallel. GPUs require a different type of coding and therefore are written in a different language. We will be using the **OpenGL Shader Language** (**GLSL**), which looks very much like C/C++. For the specifications, see `https://www.khronos.org/opengl/wiki/OpenGL_Shading_Language`.

We will be creating a shader to process vertices and a shader to process fragments. Together, these will replace all the OpenGL drawing operations we've used to date. This will give you a chance to explore the movement of mathematical operations from the CPU to the GPU while at the same time understanding how mathematics principles transcend the graphics environment in which they are applied.

The first shader type we will explore is the vertex shader. This code processes one vertex at a time while running multiple versions of itself in parallel, each processing a different vertex. The simplest vertex shader that can be written is one that can draw a single point. In OpenGL 2.x, to draw a point on the screen, the code would look like this:

```
glBegin()
glVertex3f(1.0,5.0,2.0);
glEnd()
```

This would be embedded in a fuller program that first sets out the projection and ModelView matrix. A vertex shader to achieve the same thing would look like this:

```
#version 330 core
void main()
{
    gl_Position = vec4(1.0, 5.0, 2.0, 1.0);
}
```

The first line informs the compiler of the version of OpenGL we are using. Next, the body of the vertex program is contained within a main() function. gl_Position is a predefined variable within the system that must be given a value within the main() function. You do not create it, but you must assign a value to it that is a 4D vector. The variable name must be spelled and capitalized exactly as it is written here. The reason for this is that the value of gl_Position is passed through the **geometry** and **rasterization** processes (*steps 2* and *3* in *Figure 17.1*). In other words, the geometry and rasterization processes are expecting it, so you need to ensure it has a value.

The objective of the fragment shader is to set a pixel color. Therefore, the output of the fragment shader is a color variable. A very simple fragment shader looks like this:

```
#version 330 core
out vec4 frag_color;
void main()
{
    frag_color = vec4(1, 1, 1, 1);
}
```

This example is setting the color to be white. Unlike the vertex shader, note that the output variable of the fragment shader has to be declared above the main() function, though you can name this variable anything you like.

The code for shaders is very different from Python. Indentation does not matter and is included for readability only. Statement lines must also end with a semicolon. If you've been working with Python for a long time, many shader compilation errors that you will get can be traced back to a missing semicolon. So, as you move forward, keep that in mind.

There's also a conceptual disconnect between the vertex and fragment shaders as by the time gl_Position is passed on to the fragment shader from the vertex shader, it has become a set of fragments, and instead of being one value, it is now many. Though the fragment shader does not implicitly operate on a set of fragments, all the code contained within it focuses on just one.

Again, this is a process best understood in practice, and therefore, we will return to developing our code base to work with these shaders.

Let's do it...

In this exercise, we will complete modifying the code base to work with shaders and create a simple set of vertex and fragment shaders to draw our teapot mesh. Follow the next steps:

1. Starting with the code you finished with in the previous exercise, or using the starter set of code available on GitHub at https://github.com/PacktPublishing/Mathematics-for-Game-Programming-and-Computer-Graphics/tree/main/Chapter17/starter, open up the LoadMesh.py file and make the following changes:

```
from GraphicsData import *
from Utils import *


class LoadMesh():
    def __init__(self, vao_ref, material, draw_type,
                    model_filename, texture_file="",
                    back_face_cull=False):
        self.vertices, self.uvs, self.normals,
            self.normal_ind, self.triangles =
                self.load_drawing(model_filename)
        self.coordinates =
            self.format_vertices(self.vertices,
                                    self.triangles)
        self.draw_type = draw_type
        self.vao_ref = vao_ref
        position = GraphicsData("vec3",
                                    self.coordinates)
```

```
        position.create_variable(material.program_id,
                            "position")
```

First, ensure you've removed any unnecessary imported libraries and classes from the very top. Note that `LoadMesh()` will not inherit from `Mesh3D.py` anymore, the reason being that we will just work with loaded models from now on and it will reduce the number of modifications we need to make to the code. In fact, you can remove the `Mesh3D.py` file completely.

Inside the initialization method, the loading of the model by the `load_drawing()` method is still the same, but immediately afterward, the formatting of the data is modified to allow for easier passing to the shader. Previously, OpenGL looped through the arrays of vertices, UVs, normals, and triangles to draw each vertex inside a `glBegin()`/`glEnd()` pair. Now that we are using shaders, all the vertices must be passed to the shader inside a flattened array created with a call to `glVertexAttribPointer()`, which you will see in use later in this section.

The value of `vao_ref` is also set here. This is the value we passed through from the main program script. We've not yet discussed what this is but will very shortly when we modify the `Object` class in this section. Now, let's continue modifying the `LoadMesh` class:

```
def format_vertices(self, coordinates, triangles):
    allTriangles = []
    for t in range(0, len(triangles), 3):
        allTriangles.append(
            coordinates[triangles[t]])
        allTriangles.append(
            coordinates[triangles[t + 1]])
        allTriangles.append(
            coordinates[triangles[t + 2]])
    return np.array(allTriangles, np.float32)
```

To get the vertices into a list of triangles to put into the flattened array, we must loop through the values of vertices and triangles taken from the mesh file and arrange the vertices into triangle order. This is extremely important as the shader will assume each of the three vertices it receives specifies a single triangle.

The last change is to the `draw()` method that feeds the triangles and drawing method into the shader:

```
def draw(self):
    glBindVertexArray(self.vao_ref)
```

```
        glDrawArrays(self.draw_type, 0,
                        len(self.coordinates))


    def load_drawing(self, filename):
        vertices = []
    ..
```

`glDrawArrays()` calls the shader to action. As there might be multiple shaders, the `glBindVertexArray()` call just before indicates which **Vertex Array Object (VAO)** to use for the drawing. The vao is set up in the `Object` class.

Note that no changes are necessary for the `load_drawing()` method that finalizes the LoadMesh class.

2. It's time to modify the `Object.py` file. Modify the code, like this:

```
from LoadMesh import *
from Camera import *
from Material import *
from Uniform import *



class Object:
    def __init__(self, obj_name):
        self.name = obj_name
        self.components = []
        self.material = None
        self.vao_ref = glGenVertexArrays(1)
        glBindVertexArray(self.vao_ref)
```

The `Object` class is going to hold onto the material, which holds onto the shaders and sets up the VAO reference, which contains all the vertex information that is to be passed to the shader.

Now, when a material is added as a component to an object, as we did in the `ShaderTeapot.py` script, the material is stored by the `Object` class as it requires early access to it when drawing:

```
    def add_component(self, component):
        if isinstance(component, Transform):
            self.components.insert(0, self.components)
        if isinstance(component, Material):
```

```python
        self.material = component
    self.components.append(component)
```

There are no changes required to the get_component() method:

```python
def get_component(self, class_type):
    ..
```

However, the update() method is completely rewritten, as follows:

```python
def update(self, camera: Camera, events = None):
    self.material.use()
    for c in self.components:
        if isinstance(c, Transform):
            projection = Uniform("mat4",
                                        camera.get_PPM())
            projection.find_variable(
                self.material.program_id,
                "projection_mat")
            projection.load()

            lookat = Uniform("mat4",
                                    camera.get_VM())
            lookat.find_variable(
                self.material.program_id,
                "view_mat")
            lookat.load()

            transformation = Uniform("mat4",
                                            c.get_MVM())
            transformation.find_variable(
                self.material.program_id,
                "model_mat")
            transformation.load()
```

```
elif isinstance(c, LoadMesh):
    c.draw()
```

Because all the drawing is now completed by shaders, the update for an object now has the task of sending all the coordinate space matrices through to the shader. In determining the screen position of a vertex in the old code base, we needed the projection matrix, the view matrix, and the model matrix. The exact same transformations are required for the shaders as the process of getting a vertex drawn on the screen given its model-based coordinates are exactly the same.

Here in the `update()` method, we make use of the `Uniform` class we created to format these matrices and pass them to the vertex shader. Each matrix undergoes the same processing. First, it is formatted through the construction of a uniform object. Next, the uniform object looks for a variable in the shader that will accept the data. For example, in the case of the projection matrix, this variable is called `projection_mat`. You will see where this variable is in the shader shortly. Following this, a call to the uniform object's `load()` method, created in the *Transferring processing from the CPU to the GPU* section, will push the matrix into memory for the shader to access.

3.   Lastly, we need to write a vertex and fragment shader to be used by this new code. In the project, create a subfolder called `shaders`. Inside this folder, create two text files, one called `vertexcolvert.vs` and one called `vertexcolfrag.vs`. Note that the names of these files should be exactly like those passed through to the `Material` class from `ShaderTeapot.py`. In this line, we've already added the following:

```
mat = Material("shaders/vertexcolvert.vs",
                "shaders/vertexcolfrag.vs")
```

4.   Open `vertexcolvert.vs`, and add this code:

```
#version 330 core
in vec3 position;
uniform mat4 projection_mat;
uniform mat4 view_mat;
uniform mat4 model_mat;
void main()
{
    gl_Position = projection_mat *
        transpose(view_mat) * transpose(model_mat) *
        vec4(position, 1);
}
```

The first thing you will notice is the declaration of the uniform values at the top. Note, these have the exact same spelling as the variables set in the update() method of the Object class. This is important as that's what links the Python data to the shader. The position variable is a Vector3 value representing the vertex currently being processed.

Inside the main() function, the screen position of the vertex is calculated using the same mathematics we used previously with older OpenGL calls. Because of the order of the multiplication and the fact the multiplication is happening all at once inside the shader rather than through multiple OpenGL calls as we did in the previous version of the project, some matrices need to be transposed (this means changing the rows to columns and vice versa). In addition, the camera's view matrix is inverted before being multiplied by the model matrix. The reason for this is that the camera is looking at the vertex, and however the camera moves, its opposite movement is essentially being added to the vertex. Before, we just had to apply the camera's movement to the OpenGL matrix stack, but with shaders, this doesn't happen.

5. Before continuing, the projection matrix in the Camera.py class needs a small tweak to change the order of its columns and rows. This is easily done inside the Camera class. Open Camera.py and make the following change to the projection matrix definition:

```
def __init__(self, fovy, aspect, near, far):
    f = 1/math.tan(math.radians(fovy/2))
    ..
    self.PPM = np.matrix([
        [a, 0, 0, 0],
        [0, b, 0, 0],
        [0, 0, c, d],
        [0, 0, -1, 0]
    ])
    self.VM = np.identity(4)
    ..
```

6. Let's now write up the fragment shader code before we run our new program for the first time. Open vertexcolfrag.vs, and add this code:

```
#version 330 core
out vec4 frag_color;
void main()
{
    frag_color = vec4(1, 1, 1, 1);
}
```

This code simply outputs the color white. For every required fragment that becomes a pixel, according to the geometry of the teapot mesh, the pixel will appear white.

You can now run the project to see the teapot displayed as a white mesh, as shown in *Figure 17.4*:

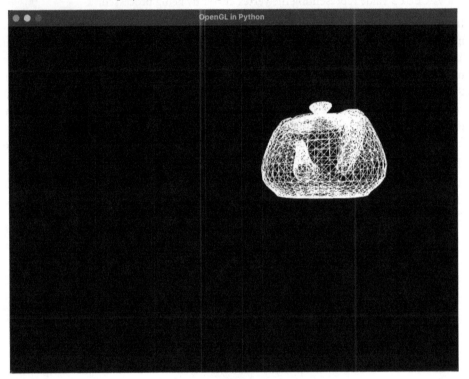

Figure 17.4: A teapot drawn with shaders

In this exercise, we completed the conversion of our code base to render graphics using purpose-built shaders. Notice that there was a lot of the code that wasn't touched, especially the transformations that underpin all the drawing and animation processes. These fundamentals don't change.

7. At this point, if you try to move the camera, you will find that the keys for moving seem to be reversed, and instead of going forward with the *w* key, you will go backward. Why? Well, we are no longer relying on OpenGL to maintain the view matrix and are setting and calculating it ourselves. In order to get objects to draw in front of the camera, we've always had to push them in the negative *z* direction. What this tells us is that the initial view is looking down the negative *z* axis, and we must initialize our camera to have a forward vector of (0, 0, -1). Instead, we've set the view matrix to an identity matrix, which assumes a forward-facing vector of (0, 0, 1). How do I figure this out?

As discussed in *Chapter 14, Working with Coordinate Spaces*, the first row in the view matrix is the camera's right axis, the second row is the up axis, and the third row is the *z* axis. But the identity matrix has (0, 0, 1, 0) in the third row, whereas we require it to be (0, 0, -1, 0) to default it to OpenGL convention. Therefore, you should now edit `Camera.py` and add the following code:

```
self.PPM = np.matrix([
    [a, 0, 0, 0],
    [0, b, 0, 0],
    [0, 0, c, d],
    [0, 0, -1, 0]
])
self.VM = np.identity(4)
self.transform = Transform()
self.transform.MVM = np.matrix([[1, 0, 0, 0],
                                [0, 1, 0, 0],
                                [0, 0, -1, 0],
                                [0, 0, 0, 1]])
self.transform.rotate_y(90, True)
self.pan_speed = 0.1
```

This will ensure the camera that we are now manually maintaining begins by facing in the correct direction. If you run the code now, you will see the keys that move the camera around are working as they did before.

If you've not worked with shaders before, the idea of them will seem a little foreign at first, though as you will see while we progress, they render objects much faster. In fact, if you were to add more teapots at this point or use the alternative teapot model with more vertices, you would notice a dramatic change in performance as you moved around the environment.

Summary

Shaders are a more efficient way of rendering as they speak directly to the graphics card. However, the way in which they process one vertex or one fragment at a time in parallel can initially be confusing and frustrating. Indeed, the mathematics used in them can appear rawer than other operations embedded in scripting languages and API calls, such as those used in the old version of OpenGL, as you, as the programmer, don't ever see the exact code inside higher-level methods. However, when placed in the shader code, there's nowhere to hide. The higher-level methods don't exist and you have to write a lot of the functionality from scratch. For example, whereas in OpenGL versions 2 and lower, calls to `gluPerspective()` would set up the projection matrix and automatically affect any meshes in the environment, with shader code you need to keep a copy of the projection matrix, maintain it, and then manually feed it through to the vertex shader where it is multiplied with each vertex.

In this chapter, we've updated the code base of the project we've been working on to work with shaders. The reason is twofold: first, shaders render faster, and they are the modern way to draw in computer graphics, and second, as we moved further into this section and examined rendering in more detail, you must have got a firmer understanding of the processes that go on under higher-level graphics and game engines. We began by reviewing the render pipeline and examining where shader code can be used to customize the rendering process, specifically looking at vertex and fragment shaders. Following this, we modified our project to update it to use shaders for rendering rather than relying on deprecated OpenGL.

In the next chapter, we will examine advanced surface rendering techniques that will require us to tap back into the work we've done on UVs and normals. As we continue, you'll start to see that although we've made a big change to the project code base, the mathematics stays relevant and consistent.

18

Customizing the Render Pipeline

In *Chapter 5*, *Let's Light It Up!*, we discussed adding color, textures, and lights to a scene. In our project, we colored vertices, applied textures, and turned on lights with single OpenGL calls.

Now in this chapter we will investigate how the color of all pixels is calculated through the use of **shaders**. When working with shaders, all the mathematics is revealed. With a few modifications to your project, by the end of this chapter you will be using shaders for the following purposes:

- Coloring and texturing mesh faces
- Turning on the lights

We will begin by grabbing the UV values that come with the OBJ model file and passing these values and a texture to a vertex and fragment shader for processing. This will allow us to color a model using an external image. Following this, because a plain image on a model will look rather flat, we'll examine the fundamental lighting models that have been traditionally used to emulate lighting in graphics. Taking the mathematics from these models, we will translate them into shader code to light up the 3D scene.

Being able to understand mathematical concepts related to the calculations used in rendering and translating them into shader code is an essential skill for today's graphics programmers. From this chapter, you will be able to examine theoretical concepts and implement them in your projects to extend the rendering possibilities.

Technical requirements

The solution files containing this chapter's code can be found on GitHub at `https://github.com/PacktPublishing/Mathematics-for-Game-Programming-and-Computer-Graphics/tree/main/Chapter1 8` in the `Chapter18` folder.

Coloring and texturing mesh faces

Thus far, in our OpenGL project, we have implemented simple white mesh rendering. Now, it's time to add the functionality of coloring and texturing polygons and pixels. The logic is similar to that of vertex coloring and UV mapping, as discussed in *Chapter 5*, *Let's Light It Up*! Though now, when using shaders to do the rendering, the colors are added vertex by vertex and pixel by pixel.

A color is allocated to a vertex via the vertex shader and then passed to the fragment shader. The color for each vertex will specify the color of the pixel representing the vertex, but not the colors in between. Take, for example, the close-up of our teapot model in *Figure 18.1*:

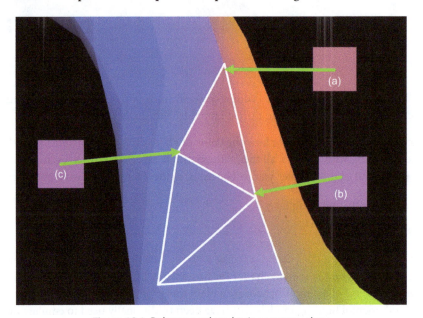

Figure 18.1: Polygons colored using vertex colors

The colors for each vector are indicated by (**a**), (**b**), and (**c**). This means the fragment shader must interpret what the value for each pixel between the vertices will be. As you can see from *Figure 18.1*, the colors fade from one vertex color to the other and across the polygon's face. This is a blending or averaging across the surface and is calculated by the graphics card. By the time the color value is used in the fragment shader, it has become an interpolated blend of the vertex colors that surround it. You do not need to do any extra calculations.

To see this process of color blending, we will modify our project so that it uses vertex colors to color each polygon in the teapot model.

Let's do it...

In this exercise, we will modify the project so that it colors the vertices of the teapot using the vertex normals. The project is already loading these normals from the mesh file, so only minimal changes need to be made:

1. Make a copy of the `Chapter_17` folder and name it `Chapter_18`.

2. First, we need to modify `LoadMesh.py` to pass the vertex normal information that's being extracted from the OBJ file to the vertex shader. Modify the code like so:

```
class LoadMesh(Mesh3D):
    def __init__(self, vao_ref, material, draw_type,
                    model_filename, texture_file="",
                    back_face_cull=False):
        self.coordinates =
                self.format_vertices(
                    self.vertices, self.triangles)
        self.normals =
            self.format_vertices(self.normals,
                                    self.normal_ind)

        ..

        position = GraphicsData("vec3",
                                    self.coordinates)
        position.create_variable(material.program_id,
                                    "position")
        vertex_normals = GraphicsData("vec3",
                                        self.normals)
        vertex_normals.create_variable(
                        material.program_id,
                        "vertex_normal")

    def format_vertices(self, coordinates, triangles):
        ..
```

We begin by reformatting the normals according to triangle order, as we had to do with the vertices in *Chapter 17, Vertex and Fragment Shading*.

Notice the new code for passing the normals to the shader is similar to that for passing the vertex positions. The difference is that the variable being created is called `vertex_normal`. This is exactly how it needs to be spelled in the vertex shader to ensure the data is passed correctly.

3. Next, we must modify `vertexcolvert.vs` so that it accepts the vertex normal value and passes it onto the fragment shader:

```
#version 330 core
in vec3 position;
in vec3 vertex_normal;
uniform mat4 projection_mat;
uniform mat4 view_mat;
uniform mat4 model_mat;
out vec3 normal;
void main()
{
    gl_Position = projection_mat *
                  inverse(transpose(view_mat)) *
                  transpose(model_mat) *
                  vec4(position, 1);
    normal = mat3(transpose(model_mat)) *
             vertex_normal;
}
```

Here, the vertex normal is presented as an incoming `vec3` value. Because this value is being passed out of the vertex shader, it needs to be given an output variable declared as `out vec3 normal`.

Inside the `main()` function, the normal is transformed into model space, which will allow it to take on transformations applied to the model.

4. Now, we can modify `vertexcolfrag.vs` so that it uses the normal color for coloring fragments:

```
#version 330 core
in vec3 normal;
out vec4 frag_color;
void main()
{
    frag_color = vec4(normal, 1);
}
```

Note that this `normal` is passed from the vertex shader as an `in` variable and then used to replace the white color we were previously using in *Chapter 17, Vertex and Fragment Shading*.

5. A few tweaks in `ShaderTeapot.py` will make the starting render easier to examine. Modify this file like this:

```
mat = Material("shaders/vertexcolvert.vs",
               "shaders/vertexcolfrag.vs")
quat_teapot.add_component(LoadMesh(
    quat_teapot.vao_ref, mat,GL_TRIANGLES,
    "models/teapot.obj"))
quat_teapot.add_component(mat)
quat_trans: Transform =
    quat_teapot.get_component(Transform)
quat_trans.update_position(pygame.Vector3(0, -1.5,
                                          -5))

quat_trans.rotate_y(90, False)

objects_3d.append(quat_teapot)
```

These changes allow the teapot to be fully rendered using `GL_TRIANGLES`. The position and rotation of the teapot have also been modified so that it appears nicely in the window when the program is run, as shown in *Figure 18.2*:

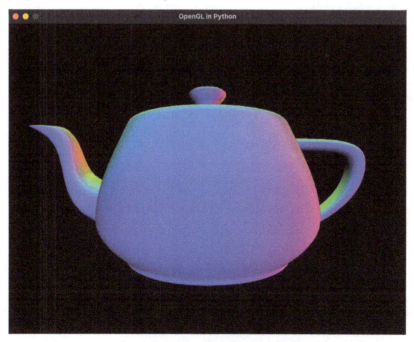

Figure 18.2: A vertex-colored teapot

6.   Run the `ShaderTeapot.py` file now to see the results.

In this exercise, you modified the project to draw a mesh as a solid object using `GL_TRIANGLES`, and also used normals to color the vertices. Next, we will modify the project and shaders to put a texture onto the teapot.

Just like the legacy version of OpenGL, shaders also require UV values to paste a texture onto the surface of a mesh. And just like vertex coloring, it's only the vertices that have the UVs specified for them. The graphics card has to perform the same calculation it does for blending color across the surface of a polygon for UV values. The result is that a texture is stretched across the polygon, ensuring the UV values specified by the vertices end up with the associated UV value location in the texture. The UV values have already been extracted with the code for the OBJ file. Now, it's time to put them into practice and texture the teapot.

Let's do it...

In this exercise, we will feed the UV values from the OBJ file and a texture through to the shader to map a surface image onto the teapot:

1.   Create a new Python script called `Texture.py`. Add the following code:

```python
import pygame
from OpenGL.GL import *


class Texture():
    def __init__(self, filename=None):
        self.surface = None
        self.texture_id = glGenTextures(1)
        if filename is not None:
            self.surface = pygame.image.load(filename)
            self.load()
```

The initialization of the class sets up a surface to hold the data from an image file, which can be in PNG or TIF format. `texture_id` has been set up as a pointer to where this texture will be in memory and used by OpenGL to access and use the pixel colors.

The `load()` method formats the data from the image file into a format suitable for use by OpenGL:

```python
    def load(self):
        width = self.surface.get_width()
```

```
height = self.surface.get_height()

pixel_data =
    pygame.image.tostring(self.surface,
                          "RGBA", 1)
glBindTexture(GL_TEXTURE_2D, self.texture_id)
glTexImage2D(GL_TEXTURE_2D, 0, GL_RGBA, width,
             height, 0, GL_RGBA,
             GL_UNSIGNED_BYTE, pixel_data)
glGenerateMipmap(GL_TEXTURE_2D)
glTexParameteri(GL_TEXTURE_2D,
                GL_TEXTURE_MAG_FILTER,
                GL_LINEAR)
glTexParameteri(GL_TEXTURE_2D,
                GL_TEXTURE_MIN_FILTER,
                GL_LINEAR_MIPMAP_LINEAR)
glTexParameteri(GL_TEXTURE_2D,
                GL_TEXTURE_WRAP_S,
                GL_REPEAT)
glTexParameteri(GL_TEXTURE_2D,
                GL_TEXTURE_WRAP_T,
                GL_REPEAT)
```

You will notice a `glBindTexture()` call, which assigns `texture_id` as the location in memory where the data will be placed. Let's explain these OpenGL calls:

- `glTexImage2D()`: Specifies the texture's size and format. Go to `https://registry.khronos.org/OpenGL-Refpages/gl4/html/glTexImage2D.xhtml` for a full description.

- `glGenerateMipmap()`: Generates mipmaps for the texture. These are a series of lesser-resolution copies of the texture that can be used when the camera is far away from the object, thus making the surface appear blurry and far away. Go to `https://registry.khronos.org/OpenGL-Refpages/gl4/html/glGenerateMipmap.xhtml` for a full description.

- `glTexParameteri()`: Sets the parameters by which OpenGL will render the texture. Go to `https://registry.khronos.org/OpenGL-Refpages/gl4/html/glTexParameter.xhtml` for a full description.

2. `LoadMesh.py` will need to be modified so that it passes a texture through to the shader:

```python
from Mesh3D import *
from GraphicsData import *
from Uniform import *
from Texture import *
from Utils import *


class LoadMesh(Mesh3D):
    def __init__(self, vao_ref, material, draw_type,
                 model_filename, texture_file="",
                 back_face_cull=False):
        self.vertices, self.uvs,
            self.uvs_ind, self.normals,
                self.normal_ind, self.triangles =
                        self.load_drawing(
                            model_filename)
        self.coordinates = self.format_vertices(
            self.vertices, self.triangles)
        self.normals = self.format_vertices(
            self.normals, self.normal_ind)
        self.uvs = self.format_vertices(
            self.uvs, self.uvs_ind)
        self.draw_type = draw_type
        self.material = material
        ..

        vertex_normals.create_variable(
            material.program_id,
            "vertex_normal")
        v_uvs = GraphicsData("vec2", self.uvs)
        v_uvs.create_variable(
            self.material.program_id,
            "vertex_uv")
        if texture_file is not None:
            self.image = Texture(texture_file)
            self.texture = Uniform("sampler2D",
```

```
                        [self.image.texture_id, 1])

    def format_vertices(self, coordinates, triangles):
..


    def draw(self):
        if self.texture is not None:
            self.texture.find_variable(
                self.material.program_id, "tex")
            self.texture.load()
        glBindVertexArray(self.vao_ref)
        glDrawArrays(self.draw_type, 0,
                    len(self.coordinates))
..
```

The Uniform class is imported to help create the sampler2D uniform type, which will pass the texture directly through to the fragment shader. You'll see where this is later in this section.

We also need to modify the model loading function in LoadMesh.py so that it returns the indices of the UVs. This will allow them to be reformatted in triangle order:

```
def load_drawing(self, filename):
    vertices = []
    uvs = []
    uvs_ind = []
    normals = []
..

    with open(filename) as fp:
        line = fp.readline()
        while line:

            ..
            if line[:2] == "f ":
                t1, t2, t3 =
                    [value for value in line[2:].split()]

..

                triangles.append(
                    [int(value) for value in
                    t3.split('/')][0] - 1)
                uvs_ind.append(
```

```
                    [int(value) for value in
                    t1.split('/')][1] - 1)
            uvs_ind.append(
                    [int(value) for value in
                    t2.split('/')][1] - 1)
            uvs_ind.append(
                    [int(value) for value in
                    t3.split('/')][1] - 1)
            normal_ind.append(
                    [int(value) for value in
                    t1.split('/')][2] - 1)
            normal_ind.append(
                    [int(value) for value in
                    t2.split('/')][2] - 1)
            normal_ind.append(
                    [int(value) for value in
                    t3.split('/')][2] - 1)
            line = fp.readline()
    return vertices, uvs, uvs_ind, normals,
        normal_ind, triangles
```

The UV values taken from the OBJ file are sent to the vertex shader in the same way as the vertices and normals were. However, the texture is treated differently. Note how it is loaded at the time of drawing, just before the vertex array.

3.   We also need a new vertex and fragment shader. Create the `texturedvert.vs` and `texturedfrag.vs` files and add the following code for `texturedvert.vs`:

```
#version 330 core
in vec3 position;
in vec3 vertex_normal;
in vec2 vertex_uv;
uniform mat4 projection_mat;
uniform mat4 model_mat;
uniform mat4 view_mat;
out vec3 normal;
out vec2 UV;
void main()
{
```

```
    gl_Position = projection_mat *
                   inverse(transpose(view_mat))
                   * transpose(model_mat) *
                   vec4(position, 1);
    UV = vertex_uv;
    normal = mat3(transpose(model_mat)) *
            vertex_normal;
}
```

In the vertex shader, the vertex UVs are simply assigned to an out variable to be sent to the fragment shader. This shader does not need to process them in any way. Here, I've left the code to calculate the normals in, even though we aren't using them right now. However, we will later on.

The code for textured frag.vs is as follows:

```
#version 330 core
out vec4 frag_color;
in vec3 normal;
in vec2 UV;
uniform sampler2D tex;

void main()
{
    frag_color = vec4(1,1,1,1);
    frag_color = frag_color * texture(tex, UV);
}
```

This code is surprisingly simple: it uses a shader function called texture(), which finds the pixel color in a texture at a specific UV coordinate. Here, I created frag_color first, making it white, and then multiplied it by the texture's pixel color.

4. Before we can run this, ShaderTeapot.py needs to be modified, like so:

```
quat_teapot = Object("Teapot")
quat_teapot.add_component(Transform())
mat = Material("shaders/texturedvert.vs",
               "shaders/texturedfrag.vs")
quat_teapot.add_component(
    LoadMesh(quat_teapot.vao_ref, mat,
            GL_TRIANGLES,
            "models/teapot.obj",
```

```
                    "images/brick.tif"))
    quat_teapot.add_component(mat)
```

The changes to the code are straightforward. The shaders that were being used have been replaced and an image to place on the teapot has been passed through to the `LoadMesh()` initialization. In this case, the `brick.tif` image file is the one I used in *Chapter 5, Let's Light It Up!* You can use the image file you used in that chapter or another one of your liking.

A note for Windows users

If you are using Windows and trying to run this, you will receive an error. Change your `frag_colour` calculation in the fragment shader to `frag_color = vec4(normal,1);`.

This will work as shown in *Figure 18.4*; I'll explain the error at the end of this exercise.

---

5. The program can now be run. The resulting textured teapot is shown in *Figure 18.3*:

Figure 18.3: The teapot, textured with a brick pattern

Why does it look so strange? Well, the UVs on the teapot weren't meant to be used with this texture, so the UVs for each vertex and image pixel will be misaligned. However, as you can see, the shader is working and placing a texture on the teapot.

Something you might like to try is replacing the following line in the fragment shader:

```
frag_color = vec4(1,1,1,1);
```

The preceding line should be replaced with the following:

```
frag_color = vec4(normal,1);
```

This will set the fragment color to the values in the normal, which is what we did in the previous exercise. This color will be multiplied by the texture pixel color and will result in the teapot shown in *Figure 18.4*:

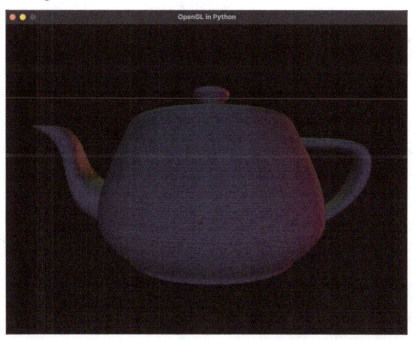

Figure 18.4: The teapot, colored with normal values and textured with a brick pattern

In this exercise, we loaded in a texture and set up UV values for a model. At this point, you know how to use any OBJ model file, along with its texture, in your projects.

Receiving shader compilation errors in Windows

While working with both the macOS and Windows platforms to create this code, I noticed that macOS is a little kinder to small issues concerning shader compilation. If you are using Windows, you will get a compilation error with the fragment shader:

```
#version 330 core
out vec4 frag_color;
in vec3 normal;
```

```
in vec2 UV;

uniform sampler2D tex;


void main()
{
    frag_color = vec4(1,1,1,1);
    frag_color = frag_color * texture(tex, UV);
}
```

Why? Because the value of in vec3 normal isn't used inside the main() function. To fix this, you must remove all mentions of normal and its use from both the vertex and fragment shader, as well as the Python script. The easiest thing to do is comment out its use in the vertex shader, like this:

```
//in vec3 vertex_normal;;
//out vec3 normal;
//normal = mat3(transpose(model_mat)) * vertex_normal;
```

Comment out its use in the fragment shader, like this:

```
//in vec3 normal;
```

Comment out its use in LoadMesh.py, like so:

```
#vertex_normals.create_variable(material.program_id,
#                                  "vertex_normal")
```

Alternatively, you could leave the normal value being used in the frag_color calculation. The only issue is that you will get a multicolored model. However, this will only happen until we discuss diffuse lighting in the *Turning on the lights* section.

Your turn...

Exercise A: Load in a properly UVed model and texture by using granny.obj and granny.png, which can be downloaded from GitHub. Modify the fragment shader and remove the normals as colors, and move and reorient the model to make it appear as follows:

Figure 18.5: A textured model

In this section, we've covered using the UVs of a model to place a texture onto that model using shader code. This is the first step in coloring a model and calculating the pixel color using the associated color in the texture file. The second step is to consider how lights in the environment affect the pixel color.

Turning on the lights

When it comes to lighting in shaders, there's a lot of hands-on mathematics. No longer can we rely on nice, neat OpenGL function calls in our main code. Instead, we must calculate the color of a pixel and consider the lighting in the fragment shader.

In the previous section, when you completed the exercise and loaded the granny model with the texture, it was unlit. It displayed the colors as they appear in the PNG texture file. To include lighting effects, we must determine how the light will influence the original color of the pixel taken from the texture. Over the years, a gradual improvement has been made to lighting models, which is evident if you take a look at the quality of computer graphics from 20 years ago up until now.

In this section, we will examine some popular lighting models and apply the mathematics to our fragment shader to light up the model.

Ambient lighting

Ambient lighting is lighting in a scene that has no apparent source. It is modeled on light that is coming from somewhere but has been scattered and bounced around the environment to provide illumination from everywhere. Think of a very cloudy day. There are no shadows but you can still see objects. The Sun's light is scattered so much through the clouds that an equal amount of light hits all surfaces.

Ambient lighting is used in computer graphics to illuminate an entire scene. The calculations are straightforward, as you are about to discover.

Let's do it...

In this exercise, we will calculate and use ambient light in our fragment shader:

1. Create a new fragment shader in your project called `lighting.vs`. Add the following code:

```
#version 330 core
out vec4 frag_color;
in vec3 normal;
in vec2 UV;
uniform sampler2D tex;

void main()
{
    vec4 lightColor = vec4(0.5, 0.5, 0.5, 1);
    float attenuation = 1;
    vec4 ambient = lightColor * attenuation;
    frag_color = texture(tex, UV) * ambient;
}
```

Here, an ambient light color of mid-gray is being set. The attenuation is the brightness of the light. To apply ambient light to the pixel color taken from the texture, it is simply multiplied.

2. In `ShaderTeapot.py` (which you may rename `ShaderGranny.py` if it makes more sense), modify the line of code that loads in the shaders so that it replaces the previous fragment shader:

```
mat = Material("shaders/texturedvert.vs",
               "shaders/lighting.vs")
```

Running this will result in a dull lit version of the model. You can also change the values of `lightColor` and `attenuation` for differing results, as shown in *Figure 18.6*:

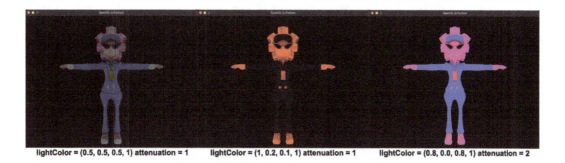

| lightColor = (0.5, 0.5, 0.5, 1) attenuation = 1 | lightColor = (1, 0.2, 0.1, 1) attenuation = 1 | lightColor = (0.8, 0.0, 0.8, 1) attenuation = 2 |

Figure 18.6: Ambient lighting effects

As ambient light does not consider the direction of the light source, there is no apparent shadowing on the object. Ambient light alone can cause a 3D model to lose its depth. This is evident in the right-most image of *Figure 18.6*, where the pink light has made the image look flat and devoid of any 3D shadowing. The simplest of the lighting models that considers the direction of the light source is the diffuse model.

Diffuse lighting

Diffuse lighting is the simplest form of lighting and is calculated using the normal vector (**N**) of the surface and the vector representing the direction of the light source (**L**), as shown in *Figure 18.7*:

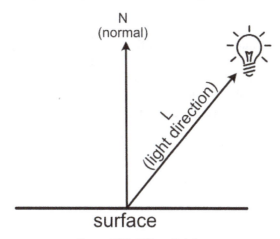

Figure 18.7: Diffuse lighting

In *Chapter 9, Practicing Vector Essentials*, we learned that we can find the angle between two vectors using the dot product operation. The light will be most intense when it is directly above the surface and the normal and light direction vectors are parallel. The light will not be reflected from the surface if it is orthogonal to the surface or below it. The dot product returns 1 when N and L are parallel, 0 when they are orthogonal, and values less than 0 when L is on the opposite side of the surface to N. The diffuse model states that the higher the value of the dot product, the brighter the surface will appear. Mathematically, we calculate this light intensity (I) as follows:

$$I = N \cdot L$$

We will now use this in the fragment shader.

Let's do it...

In this exercise, we will modify the fragment shader so that it uses the diffuse lighting model:

1. In `lighting.vs`, modify the `main()` function in the code like so:

```
void main()
{
    vec3 lightDir = vec3(-10,0,5);
    vec4 lightColor = vec4(1, 1, 1, 1);
    float NdotL = dot(normalize(normal),
                      normalize(lightDir));
    float attenuation = 1;
    vec4 diffuse = lightColor * (NdotL * attenuation);
    frag_color = texture(tex, UV) * diffuse;
}
```

Notice that a light direction has been added. This is the vector to the light source. At this point in the rendering process, you must consider that the fragment shader is working in screen space, so (0, 0, 0) is the center of the render window. The x coordinates are positive in the right direction, the y coordinates are positive in the up direction, and the z coordinates are negative going into the screen. This would place a light at 5 in the z coordinate, behind the viewer.

2. Run the program to see the effect of this diffuse light, as shown in *Figure 18.8*:

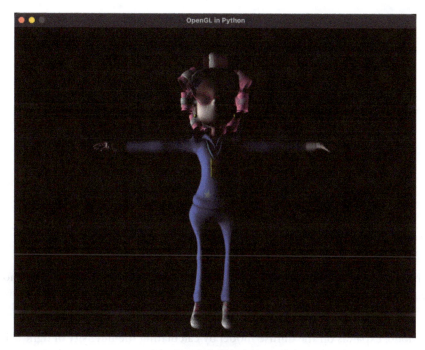

Figure 18.8: Diffuse lighting

Diffuse shading injects more depth and shadowing into a scene as it considers the direction of the light source. However, the diffuse lighting model can be improved upon by including the direction for the viewer. This is calculated in the **Phong** model.

Phong

The Phong lighting model adds specular highlights to a 3D object through the observation that an object has more gloss, shininess, or greater light reflectance based on not only the normal and light direction but also the direction of the viewer, as shown in *Figure 18.9*:

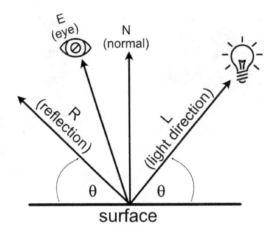

Figure 18.9: Phong lighting

When light hits a surface, it reflects off that surface at the same angle in the opposite direction. This is called the reflection vector. If the eye of the viewer were at the same angle as the reflection vector, they would experience the greatest amount of shininess.

The Phong model expands on the diffuse model by calculating the intensity of light using two factors – diffuse and **specular**. The diffuse component is calculated as it was previously; the specular component is defined as follows:

$$S = (R \cdot E)^p$$

The p variable is called the specular power. The greater the value of p, the more intense the specular highlights. The best way to experience Phong is by working it into our project.

Let's do it...

In this exercise, we will modify the lighting shader so that it includes specular highlights by implementing the Phong lighting model:

1. Modify the code in `texturedvert.vs`, like so:

    ```
    #version 330 core
    ..
    uniform mat4 view_mat;
    out vec3 normal;
    out vec2 UV;
    out vec3 view;
    void main()
    {
    ```

```
        gl_Position = projection_mat *
                    inverse(transpose(view_mat)) *
                    transpose(model_mat) *
                    vec4(position, 1);
    UV = vertex_uv;
    normal = mat3(transpose(model_mat)) *
            vertex_normal;
    view = mat3(transpose(model_mat)) * position;
}
```

Here, in the vertex shader, we are adding the view position of the vertex as it is required by the Phong model to calculate the view direction. This is calculated in world space as both the camera and model possess coordinates there.

2. Now, we must update `lighting.vs` so that we can calculate the Phong model:

```
#version 330 core
out vec4 frag_color;
in vec3 normal;
in vec2 UV;
in vec3 view;
uniform sampler2D tex;

void main()
{
    float specularPower = 2;
    vec3 lightDir = normalize(vec3(-10,0,5));
    vec3 eyeDir = normalize(-view);
    vec3 norm = normalize(normal);

    vec3 reflection =
        normalize(-reflect(lightDir,norm));
    vec4 lightColor = vec4(1, 1, 1, 1);

    //diffuse
    float NdotL = dot(norm, lightDir);
    float attenuation = 1;
    vec4 diffuse = lightColor * (NdotL * attenuation);
```

```
//specular
vec4 specular = lightColor *
                pow( dot(reflection, eyeDir),
                specularPower);


frag_color =
    texture(tex, UV) + diffuse + specular;
}
```

The majority of this code has been updated to move the normalization of all vectors to the top of the main section. Note that when working with vectors in shaders, it is best to normalize them. This is because most mathematical functions that use them assume they are normalized, and if the same vector is used multiple times, then it's more efficient to normalize them just once.

The diffuse and specular components are calculated in separate sections. Diffuse remains as it did previously. For specular, the reflection vector and direction to the viewer (eye) are required, so they are calculated. Shader language provides us with a handy `reflect()` function to perform the reflection. The direction to the eye from the fragment is calculated using *eyeDir* = *eyePosition – view*. However, when in the fragment shader, the eye position is (0, 0, 0) as we are in camera (eye) space. Hence, this calculation can be performed with `-view`.

After the specular component has been calculated, the final fragment color is devised using the pixel color from the texture and the addition of the diffuse and specular components.

3. Run your project now; you'll find a model with diffuse and shiny specular highlights, as shown in *Figure 18.10*:

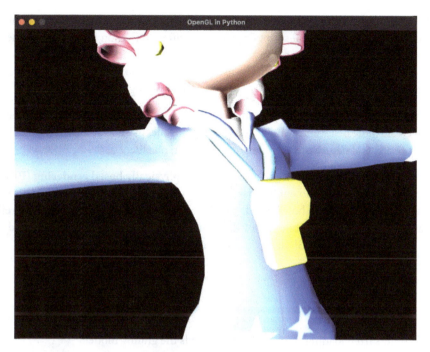

Figure 18.10: Granny light with the Phong model

In this exercise, we learned how to extend the diffuse lighting model so that it uses specular highlights. You can also add the ambient calculations here if you desire. Don't be afraid to experiment with the colors and intensities – you may just invent an entirely new shading model.

> **GLSL shading functions**
>
> There are many built-in mathematical functions in shader languages, such as the `pow()` and `reflect()` methods we've just used. As you research other shader code, you'll come across more of these functions. There are plenty of references to help you understand how each one is used. However, you can get started at `https://shaderific.com/glsl/common_functions.html` to gain an understanding of the scope of these methods.

In this section, we examined the basics of lighting a model using shaders by exploring the use of ambient, diffuse, and specular lighting. The concepts and extrapolated mathematics are deceptively simple and yet highly effective.

Summary

In this chapter, we changed over the functionality of our project, which was using older OpenGL methods, and replaced the rendering functions with our own vertex and fragment shaders. The shader code we write gets compiled into a program for the graphics processor. You will now have an understanding of how an external script or program written in Python interacts with the shader program. The majority of graphics and games programs are interactive, so being able to process transformations, projections, and camera movements in one program, and send commands to the graphics card for rendering, is a necessary skill for any graphics programmer.

The lighting models we examined here were first created by computer scientists to light scenes with 3D models to create more believable results. Though they are effective and still used today, they don't quite address the physical nature of the way light interacts in real life.

Nowadays, the more common way to light a scene is with **physically based rendering** (**PBR**). This certainly doesn't do away with ambient, diffuse, and specular lighting, but rather builds upon it by considering the nature of light and how it reacts to different surfaces as it gets absorbed and reflected. In the next and final chapter of this book, we will discuss the mathematics of PBR and build a set of vertex and fragment shaders in GLSL that you can use in your projects.

Answers

Exercise A:

The code only differs from loading the teapot since it loads the granny model:

```
quat_granny = Object("Granny")
quat_granny.add_component(Transform())
mat = Material("shaders/texturedvert.vs",
               "shaders/texturedfrag.vs")
quat_granny.add_component(LoadMesh(quat_granny.vao_ref,
    mat,GL_TRIANGLES, "models/granny.obj",
    "images/granny.png"))
quat_granny.add_component(mat)
quat_trans: Transform =
    quat_granny.get_component(Transform)
quat_trans.update_position(pygame.Vector3(0, -100, -200))
```

In the fragment shader, remove the normal from the setup of frag_color (if you aren't on Windows and haven't commented out all the error-producing code), like this:

```
frag_color = vec4(1,1,1,1)
```

19
Rendering Visual Realism Like a Pro

As graphics cards and the resolution of computer displays have improved over time, audiences have expected better quality rendering in animated movies, games, and other computer graphics-based media. By examining the way that objects in the real world get their color, computer scientists have been able to improve upon similar models of Lambert and Phong. Today, most 3D engines aimed at producing visual realism apply a shading technique called **physically based rendering** (**PBR**).

In this chapter, we will investigate the theory behind this technique and then put it into practice in our Pygame/OpenGL project. To this end, we will be discussing the following topics in this chapter:

- Following where light bounces
- Applying the Inverse Square Law
- Calculating Bidirectional Reflectance
- Putting it all together

By the end of this chapter, you will have a project that uses Python and OpenGL to render objects using PBR and understand all the components that are combined to create the final effect. This knowledge will assist you in moving forward with your independent learning of graphics to improve your own projects and also help identify improvements for others.

Technical requirements

The solution files containing the code in this chapter can be found on GitHub at `https://github.com/PacktPublishing/Mathematics-for-Game-Programming-and-Computer-Graphics/tree/main/Chapter19` in the `Chapter19` folder.

Following where light bounces

PBR is based on the actual physics of light rather than the other relatively simple lighting model of Lambert, examined in *Chapter 5, Let's Light It Up*! PBR is a concept rather than a specific algorithm and can be achieved using a variety of mathematical models. To understand how PBR works, we need to understand some key fundamentals about the visual way light works.

Light is a ray we can represent with vectors relative to the normal of the surface being hit, as shown in *Figure 19.1*:

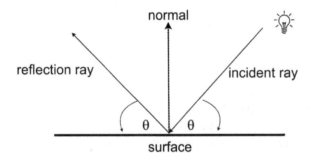

Figure 19.1: An incident and reflection ray

The light coming in from the source is called the **incident** ray and the light being reflected from the surface is called the **reflection** ray. According to the law of reflection, the angle of incidence is equal to the angle of reflection. Both rays travel in a straight line, and whether the strength of the incoming ray is the same as the reflected ray depends on what happens at the point of collision.

We also need to consider how light behaves when it passes from one medium to another. A medium can be anything, from air to metal to wood or water. Whenever the density of the medium changes, the light ray will be affected at the **point of contact**, depending on the change in density from one medium to another. The amount that light is refracted is called the **refractive index**. Some of the light will be reflected and some will pass into the medium. At the point of collision, the ray is bent in another direction. This is called **refraction**. We experience the refraction of light when looking at objects placed in water where they undergo a visual separation of themselves above the water and more dramatically when pure white light passes through a prism that splits into eight separate color components, as illustrated in *Figure 19.2*:

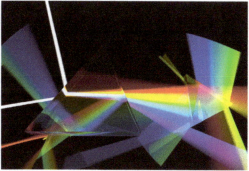

Figure 19.2: Light refraction examples

PBR enforces the principle of energy conservation, which states that *the total amount of light after hitting a surface remains the same.* Some is reflected, some is refracted, and some is absorbed.

What happens at the point of contact influences how we will see things. When light hits a mirror or highly metallic surface, almost all the light is reflected, none is refracted, and very little is absorbed. Therefore, the reflected light is almost as bright and the same color as the incoming light. Some metals absorb light at different wavelengths. For example, gold absorbs mostly blue light and reflects yellow, and hence it appears yellow.

Smooth polished objects appear shinier because of the way they reflect light. However, if you could examine such a surface on a micro level, you would see all the imperfections, as shown in *Figure 19.3*. This doesn't go unnoticed by a ray of light. These imperfections create many differing normals on the surface being hit, and therefore, the reflected ray is scattered. This dulls the appearance of the reflection. In addition, a rough surface can also make a metallic surface look less shiny because of the obvious imperfections scattering the rays, as illustrated by the brass plates in *Figure 19.3*:

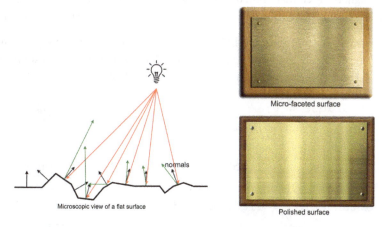

Figure 19.3: A close-up view of a surface with micro-facets and
normals and the different reflections of metal surfaces

Light rays not only bounce around an environment; their strength also diminishes over distance. This is another important factor to consider when attempting to render visual realism. We will now examine the mathematics that we can apply to calculate this effect.

Applying the Inverse Square Law

The way that the strength of light gets weaker with distance from the light source is described by the inverse square law. It states that the light intensity gets inversely weaker based on the square of the distance the viewer is away from the light source. Mathematically, we represent it like this:

$$I \propto \frac{1}{distance^2}$$

Just how quickly the light strength falls off with distance will depend on the medium through which the light is traveling. We can calculate the strength of light at a certain distance in the same medium if we know its strength for a previously measured distance. For example, if the light intensity is 10 at a distance of 100 meters from the source, we can calculate the strength that this same light will be at 125 meters, using proportions like this:

$$\frac{I_1}{I_2} = \frac{d_2^2}{d_1^2}$$

$$\frac{10}{I_2} = \frac{125^2}{100^2} = \frac{15625}{10000} = 1.5625$$

$$I_2 = 10/1.5625 = 6.4$$

This answer makes sense if we think about it as the same light at a further distance being less bright.

The strength of the light being emitted from the light source, as we discussed in *Chapter 18*, *Customizing the Render Pipeline*, is called **attenuation**, and how much it lights up a surface is known as the **radiance**.

To produce a radiance factor for numerous lights in a shader, we add the radiance of each light after determining the value of radiance, given the attenuation and distance that the light source is from the vertex or fragment. Here's some pseudo code that explains the calculations:

```
vec3 total_radiance = 0;
for(int i = 0; i < NUM_LIGHTS; ++i) //each light
{
    float distance    = length(light_data[i].position -
                                vertex_pos);
    float attenuation = light_data[i].attenuation /
                                (distance * distance);
```

```
    vec3 radiance       = light_data[i].color * attenuation;


    total_radiance += radiance;
}
```

We will put this code into our own shader soon. However, before we do, we need to consider the other influencing factors in building a PBR shader, most of which are defined by customizable reflectance functions.

Calculating Bidirectional Reflectance

Besides ordinary reflectance and scattering, PBR also integrates a **bidirectional reflectance distribution function** (**BRDF**), which considers how a specular reflection will fall off or how fuzzy it appears around the edges. It is a function that considers the four factors of the incident ray, the vector to the viewer, the surface normal, and radiance (how well the surface reflects light). In fact, the **Lambert** (diffuse) and **Phong** (specular) models we considered in *Chapter 18, Customizing the Render Pipeline*, are examples of BRDFs. The BRDF for Phong, which calculates specular lighting that can be added to the diffuse of Lambert for a final effect, can be stated as the following:

$$S = (R \cdot E)^p$$

In this formula, R is the vector of reflection of the incoming light, E is the vector from the point of contact to the viewer's eye, and p is the specular power. All vectors involved in calculating reflections are shown in *Figure 19.4*:

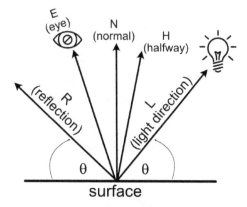

Figure 19.4: Vectors used in reflectance models

However, since the first use of Lambert, Phong, and other early reflectance models in computer graphics in the early 1970s, many updated models have been devised that consider anisotropic reflection, the Fresnel effect, and micro-facets.

One such model is the **Cook-Torrance Model** (`https://graphics.pixar.com/library/ReflectanceModel/`), which is a specular-only model considering the shininess of objects while integrating factors for micro-faceting. It takes the following form:

$$S = \frac{DFG}{4(E \cdot N)(N.L)}$$

The values of D, F, and G are the functions for a **distribution function**, the **Fresnel effect**, and **geometric attenuation** respectively. Each returns a further lighting calculation that is integrated into the final result. Basically, these three functions are plug and play – that is, there are multiple ways of calculating each one. Specifically, we will examine the work of computer graphics scientists Beckman (`https://aip.scitation.org/doi/10.1063/1.325037`), Smith (`https://ieeexplore.ieee.org/abstract/document/1138991`), and Schlick (`https://citeseerx.ist.psu.edu/viewdoc/download?doi=10.1.1.50.2297&rep=rep1&type=pdf`). We will take a look at some of these now.

Distribution functions

There are numerous distribution functions that can be used in the Cook-Torrance model. We will examine two of the most commonly used – Beckmann and GGX.

Beckmann distribution is a model used in PBR to calculate reflectance factors for micro-facets. The distribution formula results in the D value, which indicates how rough or smooth a surface is. It's described by this equation:

$$D = \frac{\exp\left(-\frac{tan^2(\alpha)}{m^2}\right)}{\pi\, m^2\, cos^4(\alpha)}, \alpha = \arccos(N \cdot H)$$

In this equation, the value of m^2 is the root mean square of the slopes of all the micro-facets. This is the average of all the gradients of the microfacet edges, each squared, and then added together. This provides the formulae with a *roughness* value between 0 and 1, where 0 indicates the surface is super smooth and 1 that it is extremely rough. If you were to manually measure all the slopes of the surface's micro-facets, squaring them, and then calculate the mean, it would be an arduous if not impossible task. Therefore, a distribution function provides us with a guesstimate of how smooth a surface is using a single roughness value.

The value of the H vector is a vector sitting halfway between the light direction and eye direction vectors shown in *Figure 19.4*. It first originated in the Blinn-Phong reflectance model, which we've not had the space to investigate herein, although if you are interested, you should follow it up here: `learnopengl.com/Advanced-Lighting/Advanced-Lighting`.

A simpler and less processor-heavy distribution function is GGX, introduced here: `www.cs.cornell.edu/~srm/publications/EGSR07-btdf.pdf`. It provides a better match to real-world reflections than Beckmann, and provides micro-faceting calculations on a number of surfaces. It is defined as the following:

$$D = \frac{a^2}{\pi((N \cdot H)^2(a^2 - 1) + 1)^2}, a = \text{roughness}^2$$

The GLSL implementation of this distribution function is as follows:

```
float GGX(float NoH, float roughness)
{
    float a = roughness*roughness;
    float a2 = a*a;
    float NoH2 = NoH*NoH;
    float numerator = a2;
    float denominator = (NoH2 * (a2 - 1.0) + 1.0);
    denominator = PI * denominator * denominator;
    return numerator / denominator;
}
```

In this code, you will find each element of the preceding equation for GGX broken down into its components. Given the dot product between the normal and halfway vector and a value for how rough the surface is, the function will return the value of D, the roughness distribution value that we can later use in calculating the Cook-Torrance BRDF.

The Fresnel effect

The Fresnel effect is the observation that the amount of reflection is highest when the viewing angle with respect to the surface normal is large. This is easily observed when examining a pool of water. If you look straight down into the water, you'll be able to see the bottom. However, when you look at it from a sharp angle, close to the water surface, it will be highly reflective, as illustrated in the photos in *Figure 19.5*:

Figure 19.5: Looking into water from different angles

When looking across the lake, the sky and the trees beyond are clearly reflected on the water's surface. However, when looking down into the water at a higher angle, there's far less reflection and it is easier to see the bottom.

Looking across a surface almost in parallel with that surface is called a **grazing angle** because you're almost grazing the surface. For a smooth surface, such as water or even smooth plastic, the reflectance tends to be very close to 100%. For rough surfaces, it is much less but reflection is still possible.

An inexpensive approximation for calculating the Fresnel term was devised by researcher Christophe Schlick and takes the optimized form:

$$F = metallicness + (1 - metallicness) * (1.0 - (H \cdot V))^5$$

The *metallic* parameter in this formula is a value between 0 and 1 that describes how close a surface is to replicating metal. For a metallic surface, the Fresnel effect is greater, as metals reflect light more readily.

The GLSL implementation of this Fresnel approximation function is as follows:

```
vec3 Fresnel(float HoV, vec3 metalness)
{
    return metalness + (1.0 - metalness) *
                pow(clamp(1.0 - HoV, 0.0, 1.0), 5.0);
}
```

The use of the `clamp` function here is to ensure (`1.0 - HoV`) does not go outside the range of 0 and 1.

Geometric attentuation factor

The geometric attenuation factor is a value that describes the self-shadowing and masking on a surface due to micro-facets. Its equation is in the following form:

$$G = \text{GS}(N \cdot V, \text{roughness}) \times GS(N \cdot L, \text{roughness})$$

And the GS function is defined as the following:

$$\text{GS(Ndot, roughness)} = \frac{Ndot}{Ndot \times \left(1.0 - \dfrac{(roughness + 1)^2}{8}\right) + \dfrac{(roughness + 1)^2}{8}}$$

It returns a single float value that is multiplied by the distribution function and Fresnel equation results to add even more visual realism to a surface.

The GLSL implementation of Smith's approximation function with Schlick's optimization, according to the preceding equations, is the following:

```
float GASchlick(float NoV, float roughness)
{
    float r = (roughness + 1.0);
    float k = (r*r) / 8.0;
    float numerator   = NoV;
    float denominator = NoV * (1.0 - k) + k;
    return numerator / denominator;
}


float GASmith(float NoV, float NoL, float roughness)
{
    float gas2  = GASchlick(NoV, roughness);
    float gas1  = GASchlick(NoL, roughness);
    return gas1 * gas2;
}
```

In this code, you will find the preceding formulae for geometric attenuation broken down into their elements to calculate a value for G.

Further references

For more functions that can be used with BRDF and further mathematical explanations, please see these excellent references:

`https://google.github.io/filament/Filament.html`

`https://learnopengl.com/PBR/Theory`

In this section, we've covered the primary mathematics required to implement a PBR shader with particular emphasis on the BRDF. Along with direct light information, coloring, and ambient lighting, we can now implement a full PBR shader into our project.

Putting it all together

PBR lighting models are used in many game engines, including Unity and Unreal. *Walt Disney Pictures* and *Pixar* also use PBR to light their 3D animations, and in fact, the models you've learned about herein are used in their graphics tools.

What distinguishes the BRDF used by PBR is that it allows for the use of parameters. These parameters allow you to customize the look of the shader and define the surface qualities of objects, using **albedo** for the diffuse color and values for **metallicness**, **roughness**, and **ambient occlusion (AO)**.

Now, it's time to put all this theory into practice, so we can see it at work in our Python/OpenGL project.

Let's do it...

In this exercise, we will rework the project to pass the settings for albedo, metallic, roughness, and ambient occlusion through to the shaders, in addition to adding multiple lights:

1. Make a copy of the `Chapter_18` folder and rename it `Chapter_19`.

2. You will need a copy of the `sphere.obj` model file available in GitHub. Make sure you add it to the `models` folder of `Chapter_19` for your project.

3. Make a copy of `ShaderTeapot.py` or the file you copied from this and displayed the Granny model in *Chapter 18, Customizing the Render Pipeline*. Call this copied file `PBR.py`. Make the following changes:

```
..
from Settings import *
from Light import *

pygame.init()
..
objects_3d = []
camera = Camera(60, (screen_width / screen_height),
                0.01, 10000.0)

for x in range(10):
    for y in range(10):
        sphere = Object("Sphere")
```

```
      sphere.add_component(Transform())
      mat = Material("shaders/pbrvert.vs",
                     "shaders/pbrfrag.vs")
      sphere.add_component(
          LoadMesh(sphere.vao_ref, mat,
                   GL_TRIANGLES,
                   "models/sphere.obj"))
      sphere_mesh: LoadMesh =
          sphere.get_component(LoadMesh)
      sphere_mesh.set_properties(
          pygame.Vector3(1, 0, 1),
          x/10.0, x/10.0, y/10.0)
      sphere.add_component(mat)
      sphere_trans: Transform =
            sphere.get_component(Transform)
      sphere_trans.update_position(
            pygame.Vector3(x*20, y*20, -20))
      objects_3d.append(sphere)
```

Here, we are adding nested loops that range the values of *x* and *y* from 0 through to 10. As the *x* and *y* values change, they are used to position the spheres and also to set the properties. The set_properties() method we are yet to add to the LoadMesh() class will allow you to set the albedo, metallic, roughness, and ambient occlusion values that each sphere will use as a customized setting sent to the shader.

This code includes a newly imported file that we are yet to write (Light.py); however, it will define the lights you are adding here:

```
lights = []
lights.append(Light(pygame.Vector3(0, 100, 200),
                    pygame.Vector3(0, 1, 1), 5, 0))
lights.append(Light(pygame.Vector3(0, 50, 200),
                    pygame.Vector3(1, 0, 1), 2, 1))
lights.append(Light(pygame.Vector3(100, 0, 200),
                    pygame.Vector3(1, 1, 0), 5, 2))
..
pygame.mouse.set_visible(False)
```

```
while not done:

    . .


    glClear(GL_COLOR_BUFFER_BIT | GL_DEPTH_BUFFER_BIT)
    camera.update()
    for o in objects_3d:
        o.update(camera, lights, events)

. .
```

You can see the lights are placed into an array, called `lights`, after a set of spheres is created from the newly added `sphere.obj` model.

The parameters passed to the creation of the lights specify their location in the world, their color, their attenuation, and the position at which they appear in the array.

4. Create a new Python script called `Light.py` and add the following:

```
from Transform import *


class Light:
    def __init__(self, position=pygame.Vector3(0, 0,
                    0),color=pygame.Vector3(1, 1, 1),
                    atten=0, light_number=0):
        self.position = position
        self.atten = atten
        self.color = color
        self.light_variable =
            "light_data[" + str(light_number) +
            "].position"
        self.atten_variable = "light_data[" +
            str(light_number) + "].attenuation"
        self.color_variable = "light_data[" +
            str(light_number) + "].color"
```

The position and color for each light are set as `Vector3`. In the initialization, three variable strings for each light are also defined. They specify what the light is called in the shader code. It is essential they are spelled here as they are in the shader. For example, if the light has a `light_number` of 2, then the `self.light_variable` string will contain `light_data[2].position`. You'll see where this goes in the shader code in *step 10*.

5. Because the lights are objects that apply to each and every object in the 3D environment, they are dealt with like the camera. Open `Object.py` and modify the code thus:

```
from LoadMesh import *

..

from Light import *


class Object:
    def __init__(self, obj_name):
        ..


    def add_component(self, component):
        ..)


    def get_component(self, class_type):
        ..


    def update(self, camera: Camera,
            lights: Light([]), events = None):
        self.material.use()
        for c in self.components:
            if isinstance(c, Transform):
                ..
                transformation.load()

                for l in lights:
                    light_pos = Uniform("vec3",
                                    l.position)
                    light_pos.find_variable(
                        self.material.program_id,
                        l.light_variable)
                    light_pos.load()
                    light_atten = Uniform("float",
                                    l.atten)
                    light_atten.find_variable(
                        self.material.program_id,
```

```
                       l.atten_variable)
            light_atten.load()
            color = Uniform("vec3", l.color)
            color.find_variable(
                 self.material.program_id,
                 l.color_variable)
            color.load()


       elif isinstance(c, LoadMesh):
            c.draw()
```

In this code, lights are passed to the update() function as an array. This array is looped over and the uniform variables in the shader for the light's position, color, and attenuation are passed through.

6. LoadMesh.py also needs a small modification, thus:

```
class LoadMesh(Mesh3D):
    def __init__(self, vao_ref, material, draw_type,
                 model_filename, texture_file="",
                 back_face_cull=False):

    . .

        #Comment out v_uvs as they aren't needed for
        #the shader and will cause Windows errors
        #v_uvs = GraphicsData("vec2", self.uv_vals)
        #v_uvs.create_variable(
        #    self.material.program_id,
        #    "vertex_uv")
        self.albedo = None
        self.metallic = None
        self.roughness = None
        self.ao = None
        #Comment out these next lines or remove them.
        #if texture_file is not None:
            #self.image = Texture(texture_file)
```

```python
        #self.texture = Uniform("sampler2D",
                        #[self.image.texture_id,
                        # 1])


    def format_vertices(self, coordinates, triangles):
        ..
    def set_properties(self, albedo, metallic,
                       roughness, ao):
        self.albedo = Uniform("vec3", albedo)
        self.metallic = Uniform("float", metallic)
        self.roughness = Uniform("float", roughness)
        self.ao = Uniform("float", ao)


    def draw(self):
        self.albedo.find_variable(
            self.material.program_id,
            "albedo")
        self.albedo.load()

        self.metallic.find_variable(
            self.material.program_id,
            "metallic")
        self.metallic.load()

        self.roughness.find_variable(
            self.material.program_id,
            "roughness")
        self.roughness.load()

        self.ao.find_variable(
            self.material.program_id, "ao")
        self.ao.load()

        glBindVertexArray(self.vao_ref)
        glDrawArrays(self.draw_type, 0,
```

```
                            len(self.coordinates))


          . .
```

The new `LoadMesh.py` script passes through the values for the albedo, metallic, roughness, and AO that are set when the spheres are created in `PBR.py` through the shader. Most of these values are floats and uniforms; therefore, our `Uniform.py` class needs to deal with float values. The albedo is a `Vector3` struct that represents the color of a pixel, and metallicness, roughness, and ambient occlusion are values between 0 and 1, representing either all of that property or none. A value of 1 for metallic would specify that the object to be rendered should be treated like a pure metal, such as gold. Roughness, when set to smooth, will give a very mirror-like finish with no micro-facets. The value of AO specifies how much of any pixel is in shadow.

7. Open `Uniform.py` and add the following code:

```
def load(self):
    if self.data_type == "vec3":
        glUniform3f(self.variable_id, self.data[0],
                    self.data[1], self.data[2])
    elif self.data_type == "float":
        glUniform1f(self.variable_id, self.data)
    elif self.data_type == "mat4":
        glUniformMatrix4fv(self.variable_id, 1,
        GL_TRUE, self.data)
```

8. Now, it's time to write the shader code. Create two files in the `shader` folder – one called `pbrvert.vs` and the other called `pbrfrag.vs`.

9. To `pbrvert.vs`, add the following:

```
#version 330 core
in vec3 position;
in vec3 vertex_normal;
uniform mat4 projection_mat;
uniform mat4 model_mat;
uniform mat4 view_mat;
out vec3 normal;
out vec3 world_pos;
out vec3 cam_pos;

void main()
```

```
{
    gl_Position = projection_mat * transpose(view_mat)
                    * transpose(model_mat) *
                    vec4(position, 1);
    normal = mat3(transpose(model_mat)) *
                vertex_normal;
    world_pos = (transpose(model_mat) *
                    vec4(position, 1)).rgb;
    cam_pos = vec3(inverse(transpose(model_mat)) *
                    vec4(view_mat[3][0],
                    view_mat[3][1],
                    view_mat[3][2],1));
}
```

This is very similar to vertex shaders we've written in the past; however, we are now passing through the `world_pos` (world position) of the vertex as well as the `cam_pos` (camera position). These will be required to calculate and deal with real-world vectors for the eye-viewing direction and lighting calculations.

10. To `pbrfrag.vs`, add the following:

```
#version 330 core
out vec4 frag_color;
in vec3 world_pos;
in vec3 normal;
in vec3 cam_pos;

// material parameters
uniform vec3   albedo;
uniform float metallic;
uniform float roughness;
uniform float ao;

struct light
{
    vec3 position;
    vec3 color;
    float attenuation;
};
```

```
#define NUM_LIGHTS 3
uniform light light_data[NUM_LIGHTS];

const float PI = 3.14159265359;

void main()
{
    vec3 N = normalize(normal);
    vec3 V = normalize(cam_pos - world_pos);

    vec3 color = vec3(0,0,0);
    for(int i = 0; i < NUM_LIGHTS; ++i) //each light
    {
        // calculate per-light radiance
        vec3 L = normalize(light_data[i].position -
                            world_pos);
        vec3 H = normalize(V + L);
        float distance    =
            length(light_data[i].position -
                    world_pos);
        float attenuation = light_data[i].attenuation
                            /
                            (distance * distance);
        vec3 radiance     = light_data[i].color *
                            light_data[i].attenuation;
        color += radiance;
    }

    color *= albedo * roughness * metallic *
            ao * normal;

    frag_color = vec4(color, 1.0);
}
```

In this code, we've now set up the values of `world_pos` and `cam_pos` to be passed from the vertex shader, and also created uniform values to accept albedo, metallic, roughness, and AO values.

Inside the `main` function, each light is looped over to add up its radiant effect on a fragment using its color, attenuation, and distance from the fragment.

The remainder of the fragment shader does not do anything special but ensures that, at this point, you can press **Play** and render something, as shown in *Figure 19.6*:

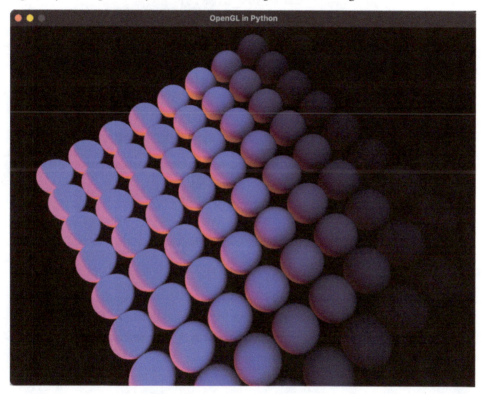

Figure 19.6: A grid of spheres with lighting effects

> **Note**
> As, at this point, we have not implemented a PBR shader, if you can't see anything in the window, try flying around with the camera to see whether you are facing it in the wrong direction.

11. It's time to modify the fragment shader to produce a PBR effect. Open `pbrfrag.vs` and make these modifications:

```glsl
#version 330 core

..


#define NUM_LIGHTS 3
uniform light light_data[NUM_LIGHTS];


const float PI = 3.14159265359;


vec3 Fresnel(float HoV, vec3 metalness)
{
    return metalness + (1.0 - metalness) *
        pow(clamp(1.0 - HoV, 0.0, 1.0), 5.0);
}


float GGX(float NoH, float roughness)
{
    float a = roughness*roughness;
    float a2 = a*a;
    float NoH2 = NoH*NoH;
    float numerator = a2;
    float denominator = (NoH2 * (a2 - 1.0) + 1.0);
    denominator = PI * denominator * denominator;
    return numerator / denominator;
}


float GASchlick(float Ndot, float roughness)
{
    float r = (roughness + 1.0);
    float k = (r*r) / 8.0;
    float numerator   = Ndot;
    float denominator = Ndot * (1.0 - k) + k;
    return numerator / denominator;
}
```

```
float GASmith(float NoV, float NoL, float roughness)
{
    float gas2  = GASchlick(NoV, roughness);
    float gas1  = GASchlick(NoL, roughness);
    return gas1 * gas2;
}
```

First, we add the functions for each of the methods required by the Cook-Torrance BDRF. These are the same functions we looked at in the *Calculating bidirectional reflectance* section.

Next, we modify the main function to use these functions and calculate the BRDF:

```
void main()
{
    vec3 N = normalize(normal);
    vec3 V = normalize(cam_pos - world_pos);

    vec3 metalness = vec3(0.01);
    metalness = mix(metalness, albedo, metallic);

    // reflectance equation
    vec3 totalRadiance = vec3(0.0);
    for(int i = 0; i < NUM_LIGHTS; ++i) //each light
    {

        ..

        float attenuation = light_data[i].attenuation
                            / (distance * distance);
        vec3 radiance     = light_data[i].color *
                    light_data[i].attenuation;

        // Cook-Torrance BRDF
        float D = GGX(max(dot(N, H), 0.0), roughness);
        float G   = GASmith(max(dot(N, V), 0.0),
                    max(dot(N, L), 0.0), roughness);
        vec3 F    = Fresnel(max(dot(H, V), 0.0),
                        metalness);
```

```
        vec3 numerator     = D * G * F;
        float denominator = 4.0 * max(dot(N, V), 0.0)
            * max(dot(N, L), 0.0) + 0.0001;
        vec3 specular      = numerator / denominator;

        // add to total radiance
        float NoL = max(dot(N, L), 0.0);
        totalRadiance += (albedo / PI + specular) *
                         radiance * NoL;
    }

    vec3 ambient = vec3(0.01) * albedo * ao;
    vec3 color = ambient + totalRadiance;

    color = color / (color + vec3(1.0));
    color = pow(color, vec3(1.0/2.2));

    frag_color = vec4(color, 1.0);
}
```

Here, the BDRF is calculated. The Cook-Torrance formula is for specular reflection only, and therefore, immediately after calculating the specular (for each light), which uses the variables for metallic and roughness, the albedo and AO are also integrated.

Running the project now, you will find a grid of spheres with differing metallic, roughness, and AO values, as shown in *Figure 19.7*:

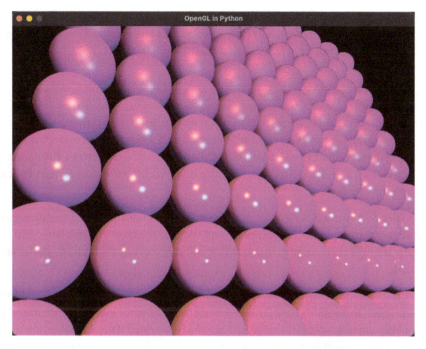

Figure 19.7: PBR of spheres with different parameters

If you'd like to examine the albedo, metallic, roughness, and AO parameter effects on individual spheres, replace the nested `for` loop in `pbr.py` to draw just one sphere with differing values, as shown in *Figure 19.8*:

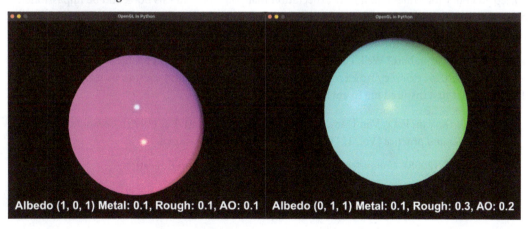

Figure 19.8: Spheres with different PBR treatments

In this exercise, you've created the first version of a PBR shader and used it to render spheres with different settings for albedo, metallicness, roughness, and ambient occlusions. As discussed, PBR is an idea for shading and not a fixed algorithm, so feel free to play with the values. I'm certain as you continue to independently investigate more shading techniques and improvements on this basic PBR shader we have created, you will be able to integrate them into your project.

Summary

As you've explored in this chapter, there's a lot of mathematics involved in creating shaders, although the basics still focus on the vectors that explain the direction of a surface with respect to the position of the light and the location of the viewer. With the addition of a few extra PBR parameters of metallicness, roughness, and AO, we are now also able to define how a surface scatters light and use that to improve a final render.

Your Python/OpenGL project is now at the point that you can continue to independently research graphics and shader techniques and experiment with them in the base that you have. You will now have a firm foundation of knowledge in this area that you can apply in the future to games and other applications alike.

The domain of mathematics involved in computer games and graphics is enormous. Unfortunately, books have page limits and authors have limited writing time. To cover everything in this field would require a set of encyclopedic volumes that would have to be updated daily. However, the underlying mathematics doesn't change. It is my hope that through reading the content herein, you will not only expand your knowledge and skills in this area but also feel confident while going forward in your own independent explorations of the content in this field, and someday add to it.

There's a never-ending list of books and online tutorials about mathematics in games and graphics to keep you busy for millennia. It's my expectation that you will read this book and be impassioned to further investigate this field, feeling confident in the skills you've obtained herein to dive into the work of others.

Where do you go from here? Well, I can recommend my own tutorials and resources at `h3dlearn.com` and `youtube.com/c/holistic3d`, although you might also want to investigate a couple of the texts that first inspired me to enter this field:

- Foley, J. D., Van, F. D., Van Dam, A., Feiner, S. K., & Hughes, J. F. (1996). *Computer graphics: principles and practice* (Vol. 12110). Addison-Wesley Professional.

- Hill Jr, F. S. (2008). *Computer graphics using OpenGL*. Pearson Education.

By giving you these older references, I'm pointing out that technology changes but the fundamentals of mathematics remain.

If you are further keen to investigate the use of shaders with Python and OpenGL, there's an excellent resource with shader code at `shadertoy.com`, and I have a YouTube series of tutorials that explain how to convert the shaders on this site for use in the code you are now familiar with: `https://youtube.com/playlist?list=PLi-ukGVOag_2FRKHY5pakPNf9b9KXaYiD`.

Wherever your mathematics, games, and graphics journey takes you from here, all my very best wishes, and I hope you develop the same passion for this field that I have.

Index

`Packt.com`

Subscribe to our online digital library for full access to over 7,000 books and videos, as well as industry leading tools to help you plan your personal development and advance your career. For more information, please visit our website.

Why subscribe?

- Spend less time learning and more time coding with practical eBooks and Videos from over 4,000 industry professionals

- Improve your learning with Skill Plans built especially for you

- Get a free eBook or video every month

- Fully searchable for easy access to vital information

- Copy and paste, print, and bookmark content

Did you know that Packt offers eBook versions of every book published, with PDF and ePub files available? You can upgrade to the eBook version at `packt.com` and as a print book customer, you are entitled to a discount on the eBook copy. Get in touch with us at `customercare@packtpub.com` for more details.

At `www.packt.com`, you can also read a collection of free technical articles, sign up for a range of free newsletters, and receive exclusive discounts and offers on Packt books and eBooks.

Other Books You May Enjoy

If you enjoyed this book, you may be interested in these other books by Packt:

Game Physics Cookbook

Gabor Szauer

ISBN: 978-1-78712-366-3

- Implement fundamental maths so you can develop solid game physics
- Use matrices to encode linear transformations
- Know how to check geometric primitives for collisions
- Build a Physics engine that can create realistic rigid body behavior
- Understand advanced techniques, including the Separating Axis Theorem
- Create physically accurate collision reactions
- Explore spatial partitioning as an acceleration structure for collisions
- Resolve rigid body collisions between primitive shapes

Applying Math with Python

Sam Morley

ISBN: 978-1-83898-975-0

- Get familiar with basic packages, tools, and libraries in Python for solving mathematical problems

- Explore various techniques that will help you to solve computational mathematical problems

- Understand the core concepts of applied mathematics and how you can apply them in computer science

- Discover how to choose the most suitable package, tool, or technique to solve a certain problem

- Implement basic mathematical plotting, change plot styles, and add labels to the plots using Matplotlib

- Get to grips with probability theory with the Bayesian inference and Markov Chain Monte Carlo (MCMC) methods

Packt is searching for authors like you

If you're interested in becoming an author for Packt, please visit authors.packtpub.com and apply today. We have worked with thousands of developers and tech professionals, just like you, to help them share their insight with the global tech community. You can make a general application, apply for a specific hot topic that we are recruiting an author for, or submit your own idea.

Hi!

I am Dr Penny de Byl, author of Mathematics for Game Programming and Computer Graphics. I really hope you enjoyed reading this book and found it useful for increasing your productivity and efficiency in Game Programming and Computer Graphics.

It would really help me (and other potential readers!) if you could leave a review on Amazon sharing your thoughts on Mathematics for Game Programming and Computer Graphics.

Go to the link below or scan the QR code to leave your review:

`https://packt.link/r/1801077339`

Your review will help me to understand what's worked well in this book, and what could be improved upon for future editions, so it really is appreciated.

Best Wishes,

Dr Penny de Byl

Download a Free PDF copy of this book

Thanks for purchasing this book!

Do you like to read on the go but are unable to carry your print books everywhere?

Is your eBook purchase not compatible with the device of your choice?

Don't worry, now with every Packt book you get a DRM-free PDF version of that book at no cost.

Read anywhere, any place, on any device. Search, copy, and paste code from your favorite technical books directly into your application.

The perks don't stop there, you can get exclusive access to discounts, newsletters, and great free content in your inbox daily

Follow these simple steps to get the benefits:

1. Scan the QR code or visit the link below

https://packt.link/free-ebook/9781801077330

2. Submit your proof of purchase
3. That's it! We'll send your free PDF and other benefits to your email directly